A Thousand-Mile Walk to the Gulf

JOHN MUIR ABOUT 1870

A Thousand-Mile Walk to the Gulf

JOHN MUIR

EDITED AND WITH
AN INTRODUCTION BY
WILLIAM FREDERIC BADÈ

FOREWORD BY
PETER JENKINS

With illustrations

A MARINER BOOK
HOUGHTON MIFFLIN COMPANY
BOSTON NEW YORK

MARINER PAPERBACKS BY JOHN MUIR

My First Summer in the Sierra

Travels in Alaska

A Thousand-Mile Walk to the Gulf

First Mariner Books edition 1998

For information about permission to reproduce selections from this book, write to trade.permissions@hmhco.com or to Permissions, Houghton Mifflin Harcourt Publishing Company, 3 Park Avenue, 19th Floor, New York, New York 10016.

www.hmhco.com

Library of Congress Cataloging-in-Publication Data is available.

ISBN-13: 978-0-395-90147-2

Printed in the United States of America

DOH 20 19 18 17

4500625980

Contents

Illustrations

Foreword

SOMETIMES I feel like going to the top of the tallest mountain and screaming, "LEAVE ME ALONE!"

It is the modern world I try so hard to keep in balance. I don't want my world to be completely wired, ordered by tiny computer chips. I don't want my face stuck in front of a computer screen day and night, my ear glued to a cell phone.

This is a major reason why I live on a farm, where the dominating sound in the spring is the singing of the birds, unless one of the local bulls is in the breeding mood and he is bellowing, daring the neighbor's bull to step across the fence. I heat part of my home with wood. Although it doesn't warm the house evenly, it allows me to know what season it is. I read books by John Muir to remind me that pure, wild nature is waiting.

I do not want to become so addicted to comforts that I do not venture out into the natural world because there is rain sometimes, and crushing heat, mud, freezing cold, insects, and trees from which bird droppings and leaves fall. I am so inspired by the wild world that I think I would become an awful person without it.

Resisting techno-creep sometimes causes me to miss out on certain things, like e-mail. For years some of the people who read my books have written me letters. The last few years, many have ended their letters with *My e-mail address is blah, blah, @ blah, blah, and I'd love to hear from you.* A few months ago I overcame my resistance and began getting e-mail and sending it.

I got e-mail from a twenty-one-year-old woman who said she was my biggest fan. She read my first book, *A Walk Across America,* at fourteen; she'd just finished reading my latest, *Along the Edge of America.* She had slapped my name into the search engine of her computer and found my e-mail address. Her dad is a geologist and her mom is a teacher in Florida, where she grew up. She now lives in Wyoming and explained that my books gave her courage to explore nature, to get to know people different from herself.

She said she loves to go to the Big Horn Mountains, that she'd spent an hour the weekend before sitting on a rock in a wilderness area watching a ground squirrel. Someday she hoped she would live in a log cabin in a valley, looking out at horses grazing in high grass.

I e-mailed her back, thanking her for reading my books. I said that there was another writer and explorer I thought she would really like, a guy named John Muir. It didn't matter that his books were written more than a hundred years ago; he was way ahead of his time.

She should begin with his first book, *A Thousand-Mile Walk to the Gulf.* Not long after the Civil War ended, Muir set out from Indiana as a very wide-eyed young man and walked to the Gulf of Mexico. I believe this walk altered the pathways he traveled for the rest of his life and led to a lifetime of exploring nature and writing about it.

John Muir was more than willing to go into parts of the world that were strange and unfamiliar to him. He said of his first experience with Florida, "I am now in the hot gardens of the sun." He seemed to know instinctively that exploring his world, seeing new creatures, feeling different winds, required the courage to feel unsure, lonely, anxious. He learned, as I have in my travels, that great rewards wait: new friends, exotic smells, and exhilarating meetings with creatures, rocks, and trees. The explorer learns that the uncertainty, insecurity, and loneliness are worth enduring.

If this young woman is to get to know John Muir, she needs to become absorbed in John Muir's world. To walk with him through Kentucky as he experiences oak trees is to be lifted high. To read of his feeling the cold air pour out of the limestone caves is to be delighted. I told her that I thought reading him would change her a little bit, help her to see things in a wind-stunted tree or the dry-heat-blasted, open plains of Wyoming in a deeper way. John Muir had a gift: he could become completely consumed by nature's full essence. He could not only appreciate a tall southern pine, a honeysuckle vine, or mountain stream but feel, see, and smell them in their fullest glory.

I assured her that no matter how much she loved "nature's grandeur" now, she would love it even more after she read John Muir. Muir was able, like few before him or since, to guide his readers into what he called the "magnificent realm."

John Muir wrote, "How narrow we selfish, conceited creatures are in our sympathies! How blind to the rights of all the rest of creation! With what dismal irreverence we speak of our fellow mortals! Though alligators, snakes, etc., naturally repel us, they are not mysterious evils. They dwell happily in these flowery wilds . . . cared for with the same species of tenderness and love as is bestowed on angels in heaven and saints on earth."

I want everyone to experience the sense of "divine harmony" that John Muir experienced so often, even if it starts as just a summer afternoon. Muir understood that once one of us experiences the "divine harmony" that exists in the wilds of nature, we will be forever changed. And we'll be forever drawn to it.

I believe we humans suffer when we are cut off from nature. I know I do. I read John Muir to be inspired, so that I will keep forging into the wild places, even if only to sit under the hundred-year-old rock maple in my yard and listen to the mourning doves over the sounds of the warm spring winds. I read him so that I will not let my desire for ease take me over. And this is why I give his books to my children, to my friends, and why I suggest to strangers like the young woman in Wyoming that she read John Muir.

"But let the children walk with nature, let them see the beautiful blendings and communions of death and life, their joyous inseparable unity, as taught in woods and meadows, plains and mountains, and streams of our blessed star."

PETER JENKINS

Introduction

"JOHN MUIR, Earth-planet, Universe."—These words are written on the inside cover of the notebook from which the contents of this volume have been taken. They reflect the mood in which the late author and explorer undertook his thousand-mile walk to the Gulf of Mexico a half-century ago. No less does this refreshingly cosmopolitan address, which might have startled any finder of the book, reveal the temper and the comprehensiveness of Mr. Muir's mind. He never was and never could be a parochial student of nature. Even at the early age of twenty-nine his eager interest in every aspect of the natural world had made him a citizen of the universe.

While this was by far the longest botanical excursion which Mr. Muir made in his earlier years, it was by no means the only one. He had botanized around the Great Lakes, in Ontario, and through parts of Wisconsin,

Indiana, and Illinois. On these expeditions he had disciplined himself to endure hardship, for his notebooks disclose the fact that he often went hungry and slept in the woods, or on the open prairies, with no cover except the clothes he wore.

"Oftentimes," he writes in some unpublished biographical notes, "I had to sleep out without blankets, and also without supper or breakfast. But usually I had no great difficulty in finding a loaf of bread in the widely scattered clearings of the farmers. With one of these big backwoods loaves I was able to wander many a long, wild mile, free as the winds in the glorious forests and bogs, gathering plants and feeding on God's abounding, inexhaustible spiritual beauty bread. Only once in my long Canada wanderings was the deep peace of the wilderness savagely broken. It happened in the maple woods about midnight, when I was cold and my fire was low. I was awakened by the awfully dismal howling of the wolves, and got up in haste to replenish the fire."

It was not, therefore, a new species of adventure upon which Mr. Muir embarked when he started on his Southern foot-tour. It was only a new response to the lure of those favorite studies which he had already pursued over uncounted miles of virgin Western forests and prairies. Indeed, had it not been for the accidental injury to his right eye in the month of March, 1867, he probably would have started somewhat earlier than he did. In a letter written to Indianapolis friends on the day after the accident, he refers mournfully to the interruption of a long-cherished plan. "For weeks," he writes, "I have daily consulted maps in locating a route through the Southern States, the West Indies, South America, and Europe — a botanical journey studied for years. And so my mind has long been in a glow with visions of the glories of a tropical flora; but, alas, I am half blind. My right eye, trained to minute analysis, is lost and I have scarce heart to open the other. Had this journey been accomplished, the stock of varied beauty acquired would have made me willing to shrink into any corner of the world, however obscure and however remote."

The injury to his eye proved to be less serious than he had at first supposed. In June he was writing to a friend: "I have been reading and botanizing for some weeks, and find that for such work I am not very much disabled. I leave this city [Indianapolis] for home to-morrow, accompanied by Merrill Moores, a little friend of mine. We will go to Decatur, Illinois, thence northward through the wide prairies, botanizing a few weeks by the way. . . . I hope to go South towards the end of the summer, and as this will be a journey that I know very little about, I hope to profit by your counsel before setting out."

In an account written after the excursion he says: "I was eager to see Illinois prairies on my way home, so we went to Decatur, near the center of the State, thence north [to Portage] by Rockford and Janesville. I botanized one week on the prairie about seven miles southwest of Pecatonica. . . . To me all plants are more precious than before. My poor eye is not better, nor worse. A cloud is over it, but in gazing over the widest landscapes, I am not always sensible of its presence."

By the end of August Mr. Muir was back again in Indianapolis. He had found it convenient to spend a "botanical week" among his University friends in Madison. So keen was his interest in plants at this time that an interval of five hours spent in Chicago was promptly turned to account in a search for them. "I did not find many plants in her tumultuous streets," he complains; "only a few grassy plants of wheat, and two or three species of weeds, — amaranth, purslane, carpet-weed, etc., — the weeds, I suppose, for man to walk upon, the wheat to feed him. I saw some green algæ, but no mosses. Some of the latter I expected to see on wet walls, and in seams on the pavements. But I suppose that the manufacturers' smoke and the terrible noise are too great for the hardiest of them. I wish I knew where I was going. Doomed to be 'carried of the spirit into the wilderness,' I suppose. I wish I could be more moderate in my desires, but I cannot, and so there is no rest."

The letter noted above was written only two days before he started on his long walk to Florida. If the concluding sentences still reflect indecision, they also convey a hint of the overmastering impulse under which he was acting. The opening sentences of his journal, afterwards crossed out, witness to this sense of inward compulsion which he felt. "Few bodies," he wrote, "are inhabited by so satisfied a soul that they are allowed exemption from extraordinary exertion through a whole life." After reciting illustrations of nature's periodicity, of the ebbs and flows of tides, and the pulsation of other forces, visible and invisible, he observes that "so also there are tides not only in the affairs of men, but in the primal thing of life itself. In some persons the impulse, being slight, is easily obeyed or overcome. But in others it is constant and cumulative in action until its power is sufficient to overmaster all impediments, and to accomplish the full measure of its demands. For many a year I have been impelled toward the Lord's tropic gardens of the South. Many influences have tended to blunt or bury this constant longing, but it has outlived and overpowered them all."

Muir's love of nature was so largely a part of his religion that he naturally chose Biblical phraseology when he sought a vehicle for his feelings. No prophet of old could have taken his call more seriously, or have entered upon his mission more frevently. During the long days of his confinement in a dark room he had opportunity for much reflection. He concluded that life was too brief and uncertain, and time too precious, to waste upon belts and saws; that while he was pottering in a wagon factory, God was making a world; and he determined that, if his eyesight was spared, he would devote the remainder of his life to a study of the process. Thus the previous bent of his habits and studies, and the sobering thoughts induced by one of the bitterest experiences of his life, combined to send him on the long journey recorded in these pages.

Some autobiographical notes found among his papers furnish interesting additional details about the period between his release from the dark room and his departure for the South. "As soon as I got out into heaven's light," he says, "I started on another long excursion, making haste with all my heart to store my mind with the Lord's beauty, and thus be ready for any fate, light or dark. And it was from this time that my long, continuous wanderings may be said to have fairly commenced. I bade adieu to mechanical inventions, determined to devote the rest of my life to the study of the inventions of God. I first went home to Wisconsin, botanizing by the way, to take leave of my father and mother, brothers and sisters, all of whom were still living near Portage. I also visited the neighbors I had known as a boy, renewed my acquaintance with them after an absence of several years, and bade each a formal good-bye. When they asked where I was going I said, 'Oh! I don't know — just anywhere in the wilderness, southward. I have already had glorious glimpses of the Wisconsin, Iowa, Michigan, Indiana, and Canada wildernesses; now I propose to go South and see something of the vegetation of the warm end of the country, and if possible to wander far enough into South America to see tropical vegetation in all its palmy glory.'

"The neighbors wished me well, advised me to be careful of my health, and reminded me that the swamps in the South were full of malaria. I stopped overnight at the home of an old Scotch lady who had long been my friend and was now particularly motherly in good wishes and advice. I told her that as I was sauntering along the road, just as the sun was going down, I heard a darling speckled-breast sparrow singing, 'The day 's done, the day 's done.' 'Weel, John, my dear laddie,' she replied, 'your day will never be done. There is no end to the kind of studies you like so well, but there's an end to mortals' strength of body and mind, to all that mortals can accomplish. You are sure to go on and on, but I want you to remember the fate of Hugh Miller.' She was one of the finest examples I ever knew of a kind, generous, great-hearted Scotchwoman."

The formal leave-taking from family and neighbors indicates his belief that he was parting from home and friends for a long time. On Sunday, the 1st of September, 1867, Mr. Muir said good-bye also to his Indianapolis friends, and went by rail to Jeffersonville, where he spent the night. The next morning he crossed the river, walked through Louisville, and struck southward through the State of Kentucky. A letter written a week later "among the hills of Bear Creek, seven miles southeast of Burkesville, Kentucky," shows that he had covered about twenty-five miles a day. "I walked from Louisville," he says, "a distance of one hundred and seventy miles, and my feet are sore. But, oh! I am paid for all my toil a thousand times over. I am in the woods on a hilltop with my back against a moss-clad log. I wish you could see my last evening's bedroom. The sun has been among the tree-tops for more than an hour; the dew is nearly all taken back, and the shade in these hill basins is creeping away into the unbroken strongholds of the grand old forests.

"I have enjoyed the trees and scenery of Kentucky exceedingly. How shall I ever tell of the miles and miles of beauty that have been flowing into me in such measure? These lofty curving ranks of lobing, swelling hills, these concealed valleys of fathomless verdure, and these lordly trees with the nursing sunlight glancing in their leaves upon the outlines of the magnificent masses of shade embosomed among their wide branches — these are cut into my memory to go with me forever.

"I was a few miles south of Louisville when I planned my journey. I spread out my map under a tree and made up my mind to go through Kentucky, Tennessee, and Georgia to Florida, thence to Cuba, thence to some part of South America; but it will be only a hasty walk. I am thankful, however, for so much. My route will be through Kingston and Madisonville, Tennessee, and through Blairsville and Gainesville, Georgia. Please write me at Gainesville. I am terribly letter-hungry. I hardly dare to think of home and friends."

In editing the journal I have endeavored, by use of all the available evidence, to trail Mr. Muir as closely as possible on maps of the sixties as well as on the most recent state and topographical maps. The one used by him has not been found, and probably is no longer in existence. Only about twenty-two towns and cities are mentioned in his journal. This constitutes a very small number when one considers the distance he covered. Evidently he was so absorbed in the plant life of the region traversed that he paid no heed to towns, and perhaps avoided them wherever possible.

The sickness which overtook him in Florida was probably of a malarial kind, although he describes it under different names. It was, no doubt, a misfortune in itself, and a severe test for his vigorous constitution. But it was also a blessing in disguise, inasmuch as it prevented him from carrying out his foolhardy plan of penetrating the tropical jungles of South America along the Andes to a tributary of the Amazon, and then floating down the river on a raft to the Atlantic. As readers of the journal will perceive, he clung to this intention even during his convalescence at Cedar Keys and in Cuba. In a letter dated the 8th of November he describes himself as "just creeping about getting plants and strength after my fever." Then he asks his correspondent to direct letters to New Orleans, Louisiana. "I shall have to go there," he writes, "for a boat to South America. I do not yet know to which point in South America I had better go." His hope to find there a boat for South America explains an otherwise mystifying letter in which he requested his brother David to send him a certain sum of money by American Express order to New Orleans. As a matter of fact he did not go into Louisiana at all, either because he learned that no south-bound ship was available at the mouth of the Mississippi, or because the unexpected appearance of the Island Belle in the harbor of Cedar Keys caused him to change his plans.

In later years Mr. Muir himself strongly disparaged the wisdom of his plans with respect to South America, as may be seen in the chapter that deals with his Cuban sojourn. The judgment there expressed was lead-penciled into his journal during a reading of it long afterwards. Nevertheless the Andes and the South American forests continued to fascinate his imagination, as his letters show, for many years after he came to California. When the long deferred journey to South America was finally made in 1911, forty-four years after the first attempt, he whimsically spoke of it as the fulfillment of those youthful dreams that moved him to undertake his thousand-mile walk to the Gulf.

Mr. Muir always recalled with gratitude the

Florida friends who nursed him through his long and serious illness. In 1898, while traveling through the South on a forest-inspection tour with his friend Charles Sprague Sargent, he took occasion to revisit the scenes of his early adventures. It may be of interest to quote some sentences from letters written at that time to his wife and to his sister Sarah. "I have been down the east side of the Florida peninsula along the Indian River," he writes, "through the palm and pine forests to Miami, and thence to Key West and the southmost keys stretching out towards Cuba. Returning, I crossed over to the west coast by Palatka to Cedar Keys, on my old track made thirty-one years ago, in search of the Hodgsons who nursed me through my long attack of fever. Mr. Hodgson died long ago, also the eldest son, with whom I used to go boating among the keys while slowly convalescing."

He then tells how he found Mrs. Hodgson and the rest of the family at Archer. They had long thought him dead and were naturally very much surprised to see him. Mrs. Hodgson was in her garden and he recognized her, though the years had altered her appearance. Let us give his own account of the meeting: "I asked her if she knew me. 'No, I don't,' she said; 'tell me your name.' 'Muir,' I replied. 'John Muir? My California John Muir?' she almost screamed. I said, 'Yes, John Muir; and you know I promised to return and visit you in about twenty-five years, and though I am a little late — six or seven years — I've done the best I could.' The eldest boy and girl remembered the stories I told them, and when they read about the Muir Glacier they felt sure it must have been named for me. I stopped at Archer about four hours, and the way we talked over old times you may imagine." From Savannah, on the same trip, he wrote: "Here is where I spent a hungry, weary, yet happy week camping in Bonaventure graveyard thirty-one years ago. Many changes, I am told, have been made in its graves and avenues of late, and how many in my life!"

Introduction

In perusing this journal the reader will miss the literary finish which Mr. Muir was accustomed to give to his later writings. This fact calls for no excuse. Not only are we dealing here with the earliest product of his pen, but with impressions and observations written down hastily during pauses in his long march. He apparently intended to use this raw material at some time for another book. If the record, as it stands, lacks finish and adornment, it also possesses the immediacy and the freshness of first impressions.

The sources which I have used in preparing this volume are threefold: (1) the original journal, of which the first half contained many interlinear revisions and expansions, and a considerable number of rough pencil sketches of plants, trees, scenery, and notable adventures; (2) a wide-spaced, typewritten, rough copy of the journal, apparently in large part dictated to a stenographer; it is only slightly revised, and comparison with the original journal shows many significant omissions and additions; (3)

two separate elaborations of his experiences in Savannah when he camped there for a week in the Bonaventure graveyard. Throughout my work upon the primary and secondary materials I was impressed with the scrupulous fidelity with which he adhered to the facts and impressions set down in the original journal.

Readers of Muir's writings need scarcely be told that this book, autobiographically, bridges the period between *The Story of my Boyhood and Youth* and *My First Summer in the Sierra.* However, one span of the bridge was lacking, for the journal ends with Mr. Muir's arrival in San Francisco about the first of April, 1868, while his first summer in the Sierra was that of 1869. By excerpting from a letter a summary account of his first visit to Yosemite, and including a description of Twenty Hill Hollow, where he spent a large part of his first year in California, the connection is made complete. The last chapter was first published as an article in the *Overland Monthly* of July, 1872.

Introduction

A revised copy of the printed article, found among Muir's literary effects, has been made the basis of the chapter on Twenty Hill Hollow as it appears in this volume.

WILLIAM FREDERIC BADÈ

A Thousand-Mile Walk to the Gulf

ROUTE OF
JOHN MUIR'S THOUSAND-MILE
WALK TO THE GULF.

NOTE:
BY RAIL FROM INDIANAPOLIS TO JEFFERSONVILLE
BY BOAT FROM SAVANNAH TO FERNANDINA.

CHAPTER I

KENTUCKY FORESTS AND CAVES

I HAD long been looking from the wild woods and gardens of the Northern States to those of the warm South, and at last, all drawbacks overcome, I set forth [from Indianapolis] on the first day of September, 1867, joyful and free, on a thousand-mile walk to the Gulf of Mexico. [The trip to Jeffersonville, on the banks of the Ohio, was made by rail.] Crossing the Ohio at Louisville [September 2], I steered through the big city by compass without speaking a word to any one. Beyond the city I found a road running southward, and after passing a scatterment of suburban cabins and cottages I reached the green woods and spread out my pocket map to rough-hew a plan for my journey.

My plan was simply to push on in a general

southward direction by the wildest, leafiest, and least trodden way I could find, promising the greatest extent of virgin forest. Folding my map, I shouldered my little bag and plant press and strode away among the old Kentucky oaks, rejoicing in splendid visions of pines and palms and tropic flowers in glorious array, not, however, without a few cold shadows of loneliness, although the great oaks seemed to spread their arms in welcome.

I have seen oaks of many species in many kinds of exposure and soil, but those of Kentucky excel in grandeur all I had ever before beheld. They are broad and dense and bright green. In the leafy bowers and caves of their long branches dwell magnificent avenues of shade, and every tree seems to be blessed with a double portion of strong exulting life. Walked twenty miles, mostly on river bottom, and found shelter in a rickety tavern.

September 3. Escaped from the dust and squalor of my garret bedroom to the glorious forest. All the streams that I tasted hereabouts

KENTUCKY OAKS

are salty and so are the wells. Salt River was nearly dry. Much of my way this forenoon was over naked limestone. After passing the level ground that extended twenty-five or thirty miles from the river I came to a region of rolling hills called Kentucky Knobs — hills of denudation, covered with trees to the top. Some of them have a few pines. For a few hours I followed the farmers' paths, but soon wandered away from roads and encountered many a tribe of twisted vines difficult to pass.

Emerging about noon from a grove of giant sunflowers, I found myself on the brink of a tumbling rocky stream [Rolling Fork]. I did not expect to find bridges on my wild ways, and at once started to ford, when a negro woman on the opposite bank earnestly called on me to wait until she could tell the "men folks" to bring me a horse — that the river was too deep and rapid to wade and that I would "sartain be drowned" if I attempted to cross. I replied that my bag and plants would ballast me; that the water did not appear to be deep, and that if I were carried away, I was a good swimmer and would soon dry in the sunshine. But the cautious old soul replied that no one ever waded that river and set off for a horse, saying that it was no trouble at all.

In a few minutes the ferry horse came gingerly down the bank through vines and weeds. His long stilt legs proved him a natural wader. He was white and the little sable negro boy that rode him looked like a bug on his back. After many a tottering halt the outward voyage was safely made, and I mounted behind little Nig. He was a queer specimen, puffy and jet as an India rubber doll and his hair was matted in sections like the wool of a merino sheep. The old horse, overladen with his black and white burden, rocked and stumbled on his stilt legs with fair promises of a fall. But all ducking signs failed and we arrived in safety among the weeds and vines of the rugged bank. A salt bath would have done us no harm. I could swim and little Afric looked as if he might float like a bladder.

I called at the homestead where my ferryman informed me I would find "tollable" water. But, like all the water of this section that I have tasted, it was intolerable with salt. Everything about this old Kentucky home bespoke plenty, unpolished and unmeasured. The house was built in true Southern style, airy, large, and with a transverse central hall that looks like a railway tunnel, and heavy rough outside chimneys. The negro quarters and other buildings are enough in number for a village, altogether an interesting representative of a genuine old Kentucky home, embosomed in orchards, corn fields and green wooded hills.

Passed gangs of woodmen engaged in felling and hewing the grand oaks for market. Fruit very abundant. Magnificent flowing hill scenery all afternoon. Walked southeast from Elizabethtown till wearied and lay down in the bushes by guess.

September 4. The sun was gilding the hilltops when I was awakened by the alarm notes of birds whose dwelling in a hazel thicket I had disturbed. They flitted excitedly close to my head, as if scolding or asking angry questions, while several beautiful plants, strangers to me, were looking me full in the face. The first botanical discovery in bed! This was one of the most delightful camp grounds, though groped for in the dark, and I lingered about it enjoying its trees and soft lights and music.

Walked ten miles of forest. Met a strange oak with willow-looking leaves. Entered a sandy stretch of black oak called "Barrens," many of which were sixty or seventy feet in height, and are said to have grown since the fires were kept off, forty years ago. The farmers hereabouts are tall, stout, happy fellows, fond of guns and horses. Enjoyed friendly chats with them. Arrived at dark in a village that seemed to be drawing its last breath. Was guided to the "tavern" by a negro who was extremely accommodating. "No trouble at all," he said.

September 5. No bird or flower or friendly tree above me this morning; only squalid garret rubbish and dust. Escaped to the woods. Came to the region of caves. At the mouth of the first I discovered, I was surprised to find ferns which belonged to the coolest nooks of Wisconsin and northward, but soon observed that each cave rim has a zone of climate peculiar to itself, and it is always cool. This cave had an opening about ten feet in diameter, and twenty-five feet perpendicular depth. A strong cold wind issued from it and I could hear the sounds of running water. A long pole was set against its walls as if intended for a ladder, but in some places it was slippery and smooth as a mast and would test the climbing powers of a monkey. The walls and rim of this natural reservoir were finely carved and flowered. Bushes leaned over it with shading leaves, and beautiful ferns and mosses were in rows and sheets on its slopes and shelves. Lingered here a long happy while, pressing specimens and printing this beauty into memory.

Arrived about noon at Munfordville; was soon discovered and examined by Mr. Munford himself, a pioneer and father of the village. He is a surveyor — has held all country offices, and every seeker of roads and lands applies to him for information. He regards all the villagers as his children, and all strangers who enter Munfordville as his own visitors. Of course he inquired my business, destination, et cetera, and invited me to his house.

After refreshing me with "parrs" he complacently covered the table with bits of rocks, plants, et cetera, things new and old which he had gathered in his surveying walks and supposed to be full of scientific interest. He informed me that all scientific men applied to him for information, and as I was a botanist, he either possessed, or ought to possess, the knowledge I was seeking, and so I received long lessons concerning roots and herbs for every mortal ill. Thanking my benefactor for his kindness, I escaped to the fields and followed a railroad along the base of a grand hill ridge. As evening came on all the dwellings I found seemed to repel me, and I could not muster courage enough to ask entertainment at any of them. Took refuge in a log schoolhouse that stood on a hillside beneath stately oaks and slept on the softest looking of the benches.

September 6. Started at the earliest bird song in hopes of seeing the great Mammoth Cave before evening. Overtook an old negro driving an ox team. Rode with him a few miles and had some interesting chat concerning war, wild fruits of the woods, et cetera. "Right heah," said he, "is where the Rebs was a-tearin' up the track, and they all a sudden thought they seed the Yankees a-comin', obah dem big hills dar, and Lo'd, how dey run." I asked him if he would like a renewal of these sad war times, when his flexible face suddenly calmed, and he said with intense earnestness, "Oh, Lo'd, want no mo wa, Lo'd no." Many of these Kentucky negroes are shrewd and intelligent, and when warmed upon a subject that interests them, are eloquent in no mean degree.

Arrived at Horse Cave, about ten miles from the great cave. The entrance is by a long easy slope of several hundred yards. It seems like a noble gateway to the birthplace of springs and fountains and the dark treasuries of the mineral kingdom. This cave is in a village [of the same name] which it supplies with an abundance of cold water, and cold air that issues from its fern-clad lips. In hot weather crowds of people sit about it in the shade of the trees that guard it. This magnificent fan is capable of cooling everybody in the town at once.

Those who live near lofty mountains may climb to cool weather in a day or two, but the overheated Kentuckians can find a patch of cool climate in almost every glen in the State. The villager who accompanied me said that Horse Cave had never been fully explored, but that it was several miles in length at least. He told me that he had never been at Mammoth Cave — that it was not worth going ten miles to see, as it was nothing but a hole in the ground, and I found that his was no rare case. He was one of the useful, practical men — too wise to waste precious time with weeds, caves, fossils, or anything else that he could not eat.

Arrived at the great Mammoth Cave. I was surprised to find it in so complete naturalness. A large hotel with fine walks and gardens is near it. But fortunately the cave has been unimproved, and were it not for the narrow trail that leads down the glen to its door, one would not know that it had been visited. There are house-rooms and halls whose entrances give but slight hint of their grandeur. And so also this magnificent hall in the mineral kingdom of Kentucky has a door comparatively small and unpromising. One might pass within a few yards of it without noticing it. A strong cool breeze issues constantly from it, creating a northern climate for the ferns that adorn its rocky front.

I never before saw Nature's grandeur in so abrupt contrast with paltry artificial gardens. The fashionable hotel grounds are in exact parlor taste, with many a beautiful plant cultivated to deformity, and arranged in strict geometrical beds, the whole pretty affair a laborious failure side by side with Divine beauty. The trees around the mouth of the cave are smooth and tall and bent forward at the bottom, then straight upwards. Only a butternut seems, by its angular knotty branches, to sympathize with and belong to the cave, with a fine growth of *Cystopteris* and *Hypnum*.

Started for Glasgow Junction. Got belated in the hill woods. Inquired my way at a farmhouse and was invited to stay overnight in a rare, hearty, hospitable manner. Engaged in familiar running talk on politics, war times, and theology. The old Kentuckian seemed to take a liking to me and advised me to stay in these hills until next spring, assuring me that I would find much to interest me in and about the Great Cave; also, that he was one of the school officials and was sure that I could obtain their school for the winter term. I sincerely thanked him for his kind plans, but pursued my own.

September 7. Left the hospitable Kentuck-

ENTRANCE TO MAMMOTH CAVE

ians with their sincere good wishes and bore away southward again through the deep green woods. In noble forests all day. Saw mistletoe for the first time. Part of the day I traveled with a Kentuckian from near Burkesville. He spoke to all the negroes he met with familiar kindly greetings, addressing them always as "Uncles" and "Aunts." All travelers one meets on these roads, white and black, male and female, travel on horseback. Glasgow is one of the few Southern towns that shows ordinary American life. At night with a well-to-do farmer.

September 8. Deep, green, bossy sea of waving, flowing hilltops. Corn and cotton and tobacco fields scattered here and there. I had imagined that a cotton field in flower was something magnificent. But cotton is a coarse, rough, straggling, unhappy looking plant, not half as good-looking as a field of Irish potatoes.

Met a great many negroes going to meeting, dressed in their Sunday best. Fat, happy looking, and contented. The scenery on approaching the Cumberland River becomes still grander. Burkesville, in beautiful location, is embosomed in a glorious array of verdant flowing hills. The Cumberland must be a happy stream. I think I could enjoy traveling with it in the midst of such beauty all my life. This evening I could find none willing to take me in, and so lay down on a hillside and fell asleep muttering praises to the happy abounding beauty of Kentucky.

September 9. Another day in the most favored province of bird and flower. Many rapid streams, flowing in beautiful flower-bordered cañons embosomed in dense woods. Am seated on a grand hill-slope that leans back against the sky like a picture. Amid the wide waves of green wood there are spots of autumnal yellow and the atmosphere, too, has the dawnings of autumn in colors and sounds. The soft light of morning falls upon ripening forests of oak and elm, walnut and hickory, and all Nature is thoughtful and calm. Kentucky is the greenest, leafiest State I have yet seen. The sea of soft temperate plant-green is deepest here.

Comparing volumes of vegetable verdure in different countries to a wedge, the thick end would be in the forests of Kentucky, the other in the lichens and mosses of the North. This verdure wedge would not be perfect in its lines. From Kentucky it would maintain its thickness long and well in passing the level forests of Indiana and Canada. From the maples and pines of Canada it would slope rapidly to the bleak Arctic hills with dwarf birches and alders; thence it would thin out in a long edge among hardy lichens and liverworts and mosses to the dwelling-places of everlasting frost. Far the grandest of all Kentucky plants are her noble oaks. They are the master existences of her exuberant forests. Here is the Eden, the paradise of oaks. Passed the Kentucky line towards evening and obtained food and shelter from a thrifty Tennessee farmer, after he had made use of all the ordinary anti-hospitable arguments of cautious comfortable families.

September 10. Escaped from a heap of uncordial kindness to the generous bosom of the woods. After a few miles of level ground in luxuriant tangles of brooding vines, I began the ascent of the Cumberland Mountains, the first real mountains that my foot ever touched or eyes beheld. The ascent was by a nearly regular zigzag slope, mostly covered up like a tunnel by overarching oaks. But there were a few openings where the glorious forest road of Kentucky was grandly seen, stretching over hill and valley, adjusted to every slope and curve by the hands of Nature — the most sublime and comprehensive picture that ever entered my eyes. Reached the summit in six or seven hours — a strangely long period of up-grade work to one accustomed only to the hillocky levels of Wisconsin and adjacent States.

CHAPTER II

CROSSING THE CUMBERLAND MOUNTAINS

I HAD climbed but a short distance when I was overtaken by a young man on horseback, who soon showed that he intended to rob me if he should find the job worth while. After he had inquired where I came from, and where I was going, he offered to carry my bag. I told him that it was so light that I did not feel it at all a burden; but he insisted and coaxed until I allowed him to carry it. As soon as he had gained possession I noticed that he gradually increased his speed, evidently trying to get far enough ahead of me to examine the contents without being observed. But I was too good a walker and runner for him to get far. At a turn of the road, after trotting his horse for about half an hour, and when he thought he was out of sight, I caught him rummaging my poor bag. Finding there only a comb, brush, towel, soap, a change of underclothing, a copy of Burns's poems, Milton's Paradise Lost, and a small New Testament, he waited for me, handed back my bag, and returned down the hill, saying that he had forgotten something.

I found splendid growths of shining-leaved *Ericaceæ* [heathworts] for which the Alleghany Mountains are noted. Also ferns of which *Osmunda cinnamomea* [Cinnamon Fern] is the largest and perhaps the most abundant. *Osmunda regalis* [Flowering Fern] is also common here, but not large. In Wood's[1] and Gray's Botany *Osmunda cinnamomea* is said to be a much larger fern than *Osmunda Claytoniana*, This I found to be true in Tennessee and southward, but in Indiana, part of Illinois, and Wisconsin the opposite is true. Found here the beautiful, sensitive *Schrankia*, or sensitive brier. It is a long, prickly, leguminous vine, with dense heads of small, yellow fragrant flowers.

[1] Alphonso Wood, *Class-book of Botany, with a Flora of the United States and Canada.* The copy of this work, carried by Mr. Muir on his wanderings, is still extant. The edition is that of 1862.

Vines growing on roadsides receive many a tormenting blow, simply because they give evidence of feeling. Sensitive people are served in the same way. But the roadside vine soon becomes less sensitive, like people getting used to teasing—Nature, in this instance, making for the comfort of flower creatures the same benevolent arrangement as for man. Thus I found that the *Schrankia* vines growing along footpaths leading to a backwoods schoolhouse were much less sensitive than those in the adjacent unfrequented woods, having learned to pay but slight attention to the tingling strokes they get from teasing scholars.

It is startling to see the pairs of pinnate leaves rising quickly out of the grass and folding themselves close in regular succession from the root to the end of the prostrate stems, ten to twenty feet in length. How little we know as yet of the life of plants — their hopes and fears, pains and enjoyments!

Traveled a few miles with an old Tennessee farmer who was much excited on account of the news he had just heard. "Three kingdoms, England, Ireland, and Russia, have declared war agin the United States. Oh, it's terrible, terrible," said he. "This big war comin' so quick after our own big fight. Well, it can't be helped, and all I have to say is, Amerricay forever, but I'd a heap rather they did n't fight."

"But are you sure the news is true?" I inquired. "Oh, yes, quite sure," he replied, "for me and some of my neighbors were down at the store last night, and Jim Smith can read, and he found out all about it in a newspaper."

Passed the poor, rickety, thrice-dead village of Jamestown, an incredibly dreary place. Toward the top of the Cumberland grade, about two hours before sundown I came to a log house, and as I had been warned that all the broad plateau of the range for forty or fifty miles was desolate, I began thus early to seek a lodging for the night. Knocking at the door, a motherly old lady replied to my request for supper and bed and breakfast, that I was welcome to the best she had, provided that I had the necessary change to pay my bill. When I told her that unfortunately I had nothing smaller than a five-dollar greenback, she said, "Well, I'm sorry, but cannot afford to keep you. Not long ago ten soldiers came across from North Carolina, and in the morning they offered a greenback that I could n't change, and so I got nothing for keeping them, which I was ill able to afford." "Very well," I said, "I'm glad you spoke of this beforehand, for I would rather go hungry than impose on your hospitality."

As I turned to leave, after bidding her good-bye, she, evidently pitying me for my tired looks, called me back and asked me if I would like a drink of milk. This I gladly accepted, thinking that perhaps I might not be successful in getting any other nourishment for a day or two. Then I inquired whether there were any more houses on the road, nearer than North Carolina, forty or fifty miles away. "Yes," she said, "it's only two miles to the next house, but beyond that there are no houses that I know of except empty ones whose owners have been killed or driven away during the war."

Arriving at the last house, my knock at the door was answered by a bright, good-natured, good-looking little woman, who in reply to my request for a night's lodging and food, said, "Oh, I guess so. I think you can stay. Come in and I'll call my husband." "But I must first warn you," I said, "that I have nothing smaller to offer you than a five-dollar bill for my entertainment. I don't want you to think that I am trying to impose on your hospitality."

She then called her husband, a blacksmith, who was at work at his forge. He came out, hammer in hand, bare-breasted, sweaty, begrimed, and covered with shaggy black hair. In reply to his wife's statement, that this young man wished to stop over night, he quickly replied, "That's all right; tell him to go into the house." He was turning to go back to his shop, when his wife added, "But he says he has n't any change to pay. He has nothing smaller than a five-dollar bill." Hesitating only a moment, he turned on his heel and said, "Tell him to go into the house. A man that comes right out like that beforehand is welcome to eat my bread."

When he came in after his hard day's work and sat down to dinner, he solemnly asked a blessing on the frugal meal, consisting solely of corn bread and bacon. Then, looking across the table at me, he said, "Young man, what are you doing down here?" I replied that I was looking at plants. "Plants? What kind of plants?" I said, "Oh, all kinds; grass, weeds, flowers, trees, mosses, ferns, — almost everything that grows is interesting to me."

"Well, young man," he queried, "you mean to say that you are not employed by the Government on some private business?" "No," I said, "I am not employed by any one except just myself. I love all kinds of plants, and I came down here to these Southern States to get acquainted with as many of them as possible."

"You look like a strong-minded man," he replied, "and surely you are able to do something better than wander over the country and look at weeds and blossoms. These are hard times, and real work is required of every man that is able. Picking up blossoms does n't seem to be a man's work at all in any kind of times."

To this I replied, "You are a believer in the Bible, are you not?" "Oh, yes." "Well, you know Solomon was a strong-minded man, and he is generally believed to have been the very wisest man the world ever saw, and yet he considered it was worth while to study plants; not only to go and pick them up as I am doing, but to study them; and you know we are told that he wrote a book about plants, not only of the great cedars of Lebanon, but of little bits of things growing in the cracks of the walls.[1]

"Therefore, you see that Solomon differed very much more from you than from me in this matter. I'll warrant you he had many a long ramble in the mountains of Judea, and had he

[1] The previously mentioned copy of Wood's Botany, used by John Muir, quotes on the title page 1 Kings iv, 33: "He spake of trees, from the cedar of Lebanon even unto the hyssop that springeth out of the wall."

been a Yankee he would likely have visited every weed in the land. And again, do you not remember that Christ told his disciples to 'consider the lilies how they grow,' and compared their beauty with Solomon in all his glory? Now, whose advice am I to take, yours or Christ's? Christ says, 'Consider the lilies.' You say, 'Don't consider them. It is n't worth while for any strong-minded man.'"

This evidently satisfied him, and he acknowledged that he had never thought of blossoms in that way before. He repeated again and again that I must be a very strong-minded man, and admitted that no doubt I was fully justified in picking up blossoms. He then told me that although the war was over, walking across the Cumberland Mountains still was far from safe on account of small bands of guerrillas who were in hiding along the roads, and earnestly entreated me to turn back and not to think of walking so far as the Gulf of Mexico until the country became quiet and orderly once more.

I replied that I had no fear, that I had but

very little to lose, and that nobody was likely to think it worth while to rob me; that, anyhow, I always had good luck. In the morning he repeated the warning and entreated me to turn back, which never for a moment interfered with my resolution to pursue my glorious walk.

September 11. Long stretch of level sandstone plateau, lightly furrowed and dimpled with shallow groove-like valleys and hills. The trees are mostly oaks, planted wide apart like those in the Wisconsin woods. A good many pine trees here and there, forty to eighty feet high, and most of the ground is covered with showy flowers. Polygalas [milkworts], solidagoes [goldenrods], and asters were especially abundant. I came to a cool clear brook every half mile or so, the banks planted with *Osmunda regalis*, *Osmunda cinnamomea*, and handsome sedges. The few larger streams were fringed with laurels and azaleas. Large areas beneath the trees are covered with formidable green briers and brambles, armed with hooked claws, and almost impenetrable. Houses are far apart and uninhabited, orchards and fences in ruins — sad marks of war.

About noon my road became dim and at last vanished among desolate fields. Lost and hungry, I knew my direction but could not keep it on account of the briers. My path was indeed strewn with flowers, but as thorny, also, as mortal ever trod. In trying to force a way through these cat-plants one is not simply clawed and pricked through all one's clothing, but caught and held fast. The toothed arching branches come down over and above you like cruel living arms, and the more you struggle the more desperately you are entangled, and your wounds deepened and multiplied. The South has plant fly-catchers. It also has plant man-catchers.

After a great deal of defensive fighting and struggling I escaped to a road and a house, but failed to find food or shelter. Towards sundown, as I was walking rapidly along a straight stretch in the road, I suddenly came in sight of ten mounted men riding abreast. They undoubtedly had seen me before I discovered them, for they had stopped their horses and were evidently watching me. I saw at once that it was useless to attempt to avoid them, for the ground thereabout was quite open. I knew that there was nothing for it but to face them fearlessly, without showing the slightest suspicion of foul play. Therefore, without halting even for a moment, I advanced rapidly with long strides as though I intended to walk through the midst of them. When I got within a rod or so I looked up in their faces and smilingly bade them "Howdy." Stopping never an instant, I turned to one side and walked around them to get on the road again, and kept on without venturing to look back or to betray the slightest fear of being robbed.

After I had gone about one hundred or one hundred and fifty yards, I ventured a quick glance back, without stopping, and saw in this flash of an eye that all the ten had turned their horses toward me and were evidently talking about me; supposedly, with reference to what my object was, where I was going, and whether it would be worth while to rob me. They all were mounted on rather scrawny horses, and all wore long hair hanging down on their shoulders. Evidently they belonged to the most irreclaimable of the guerrilla bands who, long accustomed to plunder, deplored the coming of peace. I was not followed, however, probably because the plants projecting from my plant press made them believe that I was a poor herb doctor, a common occupation in these mountain regions.

About dark I discovered, a little off the road, another house, inhabited by negroes, where I succeeded in obtaining a much needed meal of string beans, buttermilk, and corn bread. At the table I was seated in a bottomless chair, and as I became sore and heavy, I sank deeper and deeper, pressing my knees against my breast, and my mouth settled to the level of my plate. But wild hunger cares for none of these things, and my curiously compressed position prevented the too free indulgence of boisterous appetite. Of course, I was compelled to sleep with the trees in the one great bedroom of the open night.

September 12. Awoke drenched with mountain mist, which made a grand show, as it moved away before the hot sun. Passed Montgomery, a shabby village at the head of the east slope of the Cumberland Mountains. Obtained breakfast in a clean house and began the descent of the mountains. Obtained fine views of a wide, open country, and distant flanking ridges and spurs. Crossed a wide cool stream [Emory River], a branch of the Clinch River. There is nothing more eloquent in Nature than a mountain stream, and this is the first I ever saw. Its banks are luxuriantly peopled with rare and lovely flowers and overarching trees, making one of Nature's coolest and most hospitable places. Every tree, every flower, every ripple and eddy of this lovely stream seemed solemnly to feel the presence of the great Creator. Lingered in this sanctuary a long time thanking the Lord with all my heart for his goodness in allowing me to enter and enjoy it.

THE CLINCH RIVER, TENNESSEE

Discovered two ferns, *Dicksonia* and a small matted polypod on trees, common farther South. Also a species of magnolia with very large leaves and scarlet conical fruit. Near this stream I spent some joyous time in a grand rock-dwelling full of mosses, birds, and flowers. Most heavenly place I ever entered. The long narrow valleys of the mountainside, all well watered and nobly adorned with oaks, magnolias, laurels, azaleas, asters, ferns, Hypnum mosses, Madotheca [Scale-mosses], etc. Also towering clumps of beautiful hemlocks. The hemlock, judging from the common species of Canada, I regarded as the least noble of the conifers. But those of the eastern valleys of the Cumberland Mountains are as perfect in form and regal in port as the pines themselves. The latter abundant. Obtained fine glimpses from open places as I descended to the great valley between these mountains and the Unaka Mountains on the state line. Forded the Clinch, a beautiful clear stream, that knows many of the dearest mountain retreats that ever heard the music of running water. Reached Kingston before dark. Sent back my plant collections by express to my brother in Wisconsin.

September 13. Walked all day across small parallel valleys that flute the surface of the one wide valley. These flutings appear to have been formed by lateral pressure, are fertile, and contain some fine forms, though the seal of war is on all things. The roads never seem to proceed with any fixed purpose, but wander as if lost. In seeking the way to Philadelphia [in Loudon County, Tennessee], I was told by a buxom Tennessee "gal" that over the hills was much the nearer way, that she always went that way, and that surely I could travel it.

I started over the flint-ridges, but soon reached a set of enchanted little valleys among which, no matter how or in what direction I traveled, I could not get a foot nearer to Philadelphia. At last, consulting my map and compass, I neglected all directions and finally reached the house of a negro driver, with whom I put up for the night. Received a good deal of knowledge which may be of use should I ever be a negro teamster.

September 14. Philadelphia is a very filthy village in a beautiful situation. More or less of pine. Black oak most abundant. *Polypodium hexagonopterum* and *Aspidium acrostichoides* [Christmas Fern] most abundant of ferns and most generally distributed. *Osmunda claytoniana* rare, not in fruit, small. *Dicksonia* abundant, after leaving the Cumberland Mountains. *Asplenium ebeneum* [Ebony Spleenwort] quite common in Tennessee and many parts of Kentucky. *Cystopteris* [Bladder Fern], and *Asplenium filix-fœmina* not common through the same range. *Pteris aquilina* [Common Brake] abundant, but small.

Walked through many a leafy valley, shady grove, and cool brooklet. Reached Madisonville, a brisk village. Came in full view of the Unaka Mountains, a magnificent sight. Stayed over night with a pleasant young farmer.

September 15. Most glorious billowy mountain scenery. Made many a halt at open places to take breath and to admire. The road, in many places cut into the rock, goes winding about among the knobs and gorges. Dense growth of asters, liatris,[1] and grapevines.

Reached a house before night, and asked leave to stop. "Well, you're welcome to stop," said the mountaineer, "if you think you can live till morning on what I have to live on all the time." Found the old gentleman very communicative. Was favored with long "bar" stories, deer hunts, etc., and in the morning was pressed to stay a day or two.

September 16. "I will take you," said he, "to the highest ridge in the country, where you can see both ways. You will have a view of all the world on one side of the mountains and all creation on the other. Besides, you, who are traveling for curiosity and wonder,

[1] Wood's Botany, edition of 1862, furnishes the following interesting comment on *Liatris odoratissima* (Willd.), popularly known as Vanilla Plant or Deer's Tongue: "The fleshy leaves exhale a rich fragrance even for years after they are dry, and are therefore by the southern planters largely mixed with their cured tobacco, to impart its fragrance to that nauseous weed."

ought to see our gold mines. I agreed to stay and went to the mines. Gold is found in small quantities throughout the Alleghanies, and many farmers work at mining a few weeks or months every year when their time is not more valuable for other pursuits. In this neighborhood miners are earning from half a dollar to two dollars a day. There are several large quartz mills not far from here. Common labor is worth ten dollars a month.

September 17. Spent the day in botanizing, blacksmithing, and examining a grist mill. Grist mills, in the less settled parts of Tennessee and North Carolina, are remarkably simple affairs. A small stone, that a man might carry under his arm, is fastened to the vertical shaft of a little home-made, boyish-looking, back-action water-wheel, which, with a hopper and a box to receive the meal, is the whole affair. The walls of the mill are of undressed poles cut from seedling trees and there is no floor, as lumber is dear. No dam is built. The water is conveyed along some hillside until sufficient fall is obtained, a thing easily done in the mountains.

On Sundays you may see wild, unshorn, uncombed men coming out of the woods, each with a bag of corn on his back. From a peck to a bushel is a common grist. They go to the mill along verdant footpaths, winding up and down over hill and valley, and crossing many a rhododendron glen. The flowers and shining leaves brush against their shoulders and knees, occasionally knocking off their coon-skin caps. The first arrived throws his corn into the hopper, turns on the water, and goes to the house. After chatting and smoking he returns to see if his grist is done. Should the stones run empty for an hour or two, it does no harm.

This is a fair average in equipment and capacity of a score of mills that I saw in Tennessee. This one was built by John Vohn, who claimed that he could make it grind twenty bushels a day. But since it fell into other hands it can be made to grind only ten per day. All the machines of Kentucky and Tennessee are far behind the age. There is scarce a trace of that restless spirit of speculation and invention so characteristic of the North. But one way of doing things obtains here, as if laws had been passed making attempts at improvement a crime. Spinning and weaving are done in every one of these mountain cabins wherever the least pretensions are made to thrift and economy. The practice of these ancient arts they deem marks of advancement rather than of backwardness. "There's a place back heah," said my worthy entertainer, "whar there's a mill-house, an' a store-house, an' a still-house, an' a spring-house, an' a blacksmith shop — all in the same yard! Cows too, an' heaps of big gals a-milkin' them."

This is the most primitive country I have seen, primitive in everything. The remotest hidden parts of Wisconsin are far in advance of the mountain regions of Tennessee and North Carolina. But my host speaks of the "old-fashioned unenlightened times," like a philosopher in the best light of civilization. "I believe in Providence," said he. "Our fathers came into these valleys, got the richest of them, and skimmed off the cream of the soil. The worn-out ground won't yield no roastin' ears now. But the Lord foresaw this state of affairs, and prepared something else for us. And what is it? Why, He meant us to bust open these copper mines and gold mines, so that we may have money to buy the corn that we cannot raise." A most profound observation.

September 18. Up the mountain on the state line. The scenery is far grander than any I ever before beheld. The view extends from the Cumberland Mountains on the north far into Georgia and North Carolina to the south, an area of about five thousand square miles. Such an ocean of wooded, waving, swelling mountain beauty and grandeur is not to be described. Countless forest-clad hills, side by side in rows and groups, seemed to be enjoying the rich sunshine and remaining motionless only because they were so eagerly absorbing it. All were united by curves and slopes of inimitable softness and beauty. Oh, these forest gardens of our Father! What perfection, what divinity, in their architecture! What simplicity and mysterious complexity of detail! Who shall read the teaching of these sylvan pages, the glad brotherhood of rills that sing in the valleys, and all the happy creatures that dwell in them under the tender keeping of a Father's care?

September 19. Received another solemn warning of dangers on my way through the mountains. Was told by my worthy entertainer of a wondrous gap in the mountains which he advised me to see. "It is called Track Gap," said he, "from the great number of tracks in the rocks — bird tracks, bar tracks, hoss tracks, men tracks, all in the solid rock as if it had been mud." Bidding farewell to my worthy mountaineer and all his comfortable wonders, I pursued my way to the South.

As I was leaving, he repeated the warnings of danger ahead, saying that there were a good many people living like wild beasts on whatever they could steal, and that murders were sometimes committed for four or five dollars, and even less. While stopping with him I noticed that a man came regularly after dark to the house for his supper. He was armed with a gun, a pistol, and a long knife. My host told me that this man was at feud with one of his neighbors, and that they were prepared to shoot one another at sight. That neither of them could do any regular work or sleep in the same place two nights in succession. That they visited houses only for food, and as soon as the one that I saw had got his supper he went out and slept in the woods, without of course making a fire. His enemy did the same.

My entertainer told me that he was trying to make peace between these two men, because they both were good men, and if they would agree to stop their quarrel, they could then both go to work. Most of the food in this house was coffee without sugar, corn bread, and sometimes bacon. But the coffee was the greatest luxury which these people knew. The only way of obtaining it was by selling skins, or, in particular, "sang," that is ginseng,[1] which found a market in far-off China.

My path all to-day led me along the leafy banks of the Hiwassee,[2] a most impressive mountain river. Its channel is very rough, as it crosses the edges of upturned rock strata, some of them standing at right angles, or glancing off obliquely to right and left. Thus a multitude of short, resounding cataracts are produced, and the river is restrained from the headlong speed due to its volume and the inclination of its bed.

All the larger streams of uncultivated countries are mysteriously charming and beautiful, whether flowing in mountains or through swamps and plains. Their channels are inter-

[1] Muir's journal contains the following additional note: "M. County produces $5000 worth a year of ginseng root, valued at seventy cents a pound. Under the law it is not allowed to be gathered until the first of September."

[2] In his journal Muir spells the name "Hiawassee," a form which occurs on many of the older maps. The name probably is derived from the Cherokee Indian "Ayuhwasi," a name applied to several of their former settlements.

estingly sculptured, far more so than the grandest architectural works of man. The finest of the forests are usually found along their banks, and in the multitude of falls and rapids the wilderness finds a voice. Such a river is the Hiwassee, with its surface broken to a thousand sparkling gems, and its forest walls vine-draped and flowery as Eden. And how fine the songs it sings!

In Murphy [North Carolina] I was hailed by the sheriff who could not determine by my colors and rigging to what country or craft I belonged. Since the war, every other stranger in these lonely parts is supposed to be a criminal, and all are objects of curiosity or apprehensive concern. After a few minutes' conversation with this chief man of Murphy I was pronounced harmless, and invited to his house, where for the first time since leaving home I found a house decked with flowers and vines, clean within and without, and stamped with the comforts of culture and refinement in all its arrangements. Striking contrast to the uncouth transitionist establishments from the wigwams of savages to the clumsy but clean log castle of the thrifty pioneer.

September 20. All day among the groves and gorges of Murphy with Mr. Beale. Was shown the site of Camp Butler where General Scott had his headquarters when he removed the Cherokee Indians to a new home in the West. Found a number of rare and strange plants on the rocky banks of the river Hiwassee. In the afternoon, from the summit of a commanding ridge, I obtained a magnificent view of blue, softly curved mountain scenery. Among the trees I saw *Ilex* [Holly] for the first time. Mr. Beale informed me that the paleness of most of the women in his neighborhood, and the mountains in general hereabouts, was caused chiefly by smoking and by what is called "dipping." I had never even heard of dipping. The term simply describes the application of snuff to the gum by means of a small swab.

September 21. Most luxuriant forest. Many brooks running across the road. Blairsville

[Georgia], which I passed in the forenoon, seems a shapeless and insignificant village, but grandly encircled with banded hills. At night I was cordially received by a farmer whose wife, though smart and neat in her appearance, was an inveterate smoker.

September 22. Hills becoming small, sparsely covered with soil. They are called "knob land" and are cultivated, or scratched, with a kind of one-tooth cultivator. Every rain robs them of their fertility, while the bottoms are of course correspondingly enriched. About noon I reached the last mountain summit on my way to the sea. It is called the Blue Ridge and before it lies a prospect very different from any I had passed, namely, a vast uniform expanse of dark pine woods, extending to the sea; an impressive view at any time and under any circumstances, but particularly so to one emerging from the mountains.

Traveled in the wake of three poor but merry mountaineers—an old woman, a young woman, and a young man—who sat, leaned, and lay in the box of a shackly wagon that seemed to be held together by spiritualism, and was kept in agitation by a very large and a very small mule. In going down hill the looseness of the harness and the joints of the wagon allowed the mules to back nearly out of sight beneath the box, and the three who occupied it were slid against the front boards in a heap over the mules' ears. Before they could unravel their limbs from this unmannerly and impolite disorder, a new ridge in the road frequently tilted them with a swish and a bump against the back boards in a mixing that was still more grotesque.

I expected to see man, women, and mules mingled in piebald ruin at the bottom of some rocky hollow, but they seemed to have full confidence in the back board and front board of the wagon-box. So they continued to slide comfortably up and down, from end to end, in slippery obedience to the law of gravitation, as the grades demanded. Where the jolting was moderate, they engaged in conversation on love, marriage, and camp-meeting, according to the custom of the country. The old lady, through all the vicissitudes of the transportation, held a bouquet of French marigolds.

The hillsides hereabouts were bearing a fine harvest of asters. Reached Mount Yonah in the evening. Had a long conversation with an old Methodist slaveholder and mine owner. Was hospitably refreshed with a drink of fine cider.

CHAPTER III

THROUGH THE RIVER COUNTRY OF GEORGIA

SEPTEMBER 23. Am now fairly out of the mountains. Thus far the climate has not changed in any marked degree, the decrease in latitude being balanced by the increase in altitude. These mountains are highways on which northern plants may extend their colonies southward. The plants of the North and of the South have many minor places of meeting along the way I have traveled; but it is here on the southern slope of the Alleghanies that the greatest number of hardy, enterprising representatives of the two climates are assembled.

Passed the comfortable, finely shaded little town of Gainesville. The Chattahoochee River is richly embanked with massive, bossy, dark green water oaks, and wreathed with a dense growth of muscadine grapevines, whose ornate foliage, so well adapted to bank embroidery, was enriched with other interweaving species of vines and brightly colored flowers. This is the first truly southern stream I have met.

At night I reached the home of a young man with whom I had worked in Indiana, Mr. Prater. He was down here on a visit to his father and mother. This was a plain backwoods family, living out of sight among knobby timbered hillocks not far from the river. The evening was passed in mixed conversation on southern and northern generalities.

September 24. Spent this day with Mr. Prater sailing on the Chattahoochee, feasting on grapes that had dropped from the overhanging vines. This remarkable species of wild grape has a stout stem, sometimes five or six inches in diameter, smooth bark and hard wood, quite unlike any other wild or cultivated grapevine that I have seen. The grapes are very large, some of them nearly an inch in diameter, globular and fine flavored. Usually there are but three or four berries in a cluster, and when mature they drop off instead of decaying on the vine. Those which fall into the river are often found in large quantities in the eddies along the bank, where they are collected by men in boats and sometimes made into wine. I think another name for this grape is the Scuppernong,[1] though called "muscadine" here.

Besides sailing on the river, we had a long walk among the plant bowers and tangles of the Chattahoochee bottom lands.

September 25. Bade good-bye to this friendly family. Mr. Prater accompanied me a short distance from the house and warned me over and over again to be on the outlook for rattlesnakes. They are now leaving the damp lowlands, he told me, so that the danger is much greater because they are on their travels. Thus warned, I set out for Savannah, but got lost in the vine-fenced hills and hollows of the river

[1] The old Indian name for the southern species of fox-grape, *Vitis rotundifolia*, which Muir describes here. Wood's Botany listed it as *Vitis vulpina* L. and remarks, "The variety called 'Scuppernong' is quite common in southern gardens."

bottom. Was unable to find the ford to which I had been directed by Mr. Prater.

I then determined to push on southward regardless of roads and fords. After repeated failures I succeeded in finding a place on the river bank where I could force my way into the stream through the vine-tangles. I succeeded in crossing the river by wading and swimming, careless of wetting, knowing that I would soon dry in the hot sunshine.

Out near the middle of the river I found great difficulty in resisting the rapid current. Though I braced myself with a stout stick, I was at length carried away in spite of all my efforts. But I succeeded in swimming to the shallows on the farther side, luckily caught hold of a rock, and after a rest swam and waded ashore. Dragging myself up the steep bank by the overhanging vines, I spread out myself, my paper money, and my plants to dry.

Debated with myself whether to proceed down the river valley until I could buy a boat, or lumber to make one, for a sail instead of a march through Georgia. I was intoxicated with the beauty of these glorious river banks, which I fancied might increase in grandeur as I approached the sea. But I finally concluded that such a pleasure sail would be less profitable than a walk, and so sauntered on southward as soon as I was dry. Rattlesnakes abundant. Lodged at a farmhouse. Found a few tropical plants in the garden.

Cotton is the principal crop hereabouts, and picking is now going on merrily. Only the lower bolls are now ripe. Those higher on the plants are green and unopened. Higher still, there are buds and flowers, some of which, if the plants be thrifty and the season favorable, will continue to produce ripe bolls until January.

The negroes are easy-going and merry, making a great deal of noise and doing little work. One energetic white man, working with a will, would easily pick as much cotton as half a dozen Sambos and Sallies. The forest here is almost entirely made up of dim-green, knotty, sparsely planted pines. The soil is mostly white, fine-grained sand.

September 26. Reached Athens in the afternoon, a remarkably beautiful and aristocratic town, containing many classic and magnificent mansions of wealthy planters, who formerly owned large negro-stocked plantations in the best cotton and sugar regions farther south. Unmistakable marks of culture and refinement, as well as wealth, were everywhere apparent. This is the most beautiful town I have seen on the journey, so far, and the only one in the South that I would like to revisit.

The negroes here have been well trained and are extremely polite. When they come in sight of a white man on the road, off go their hats, even at a distance of forty or fifty yards, and they walk bare-headed until he is out of sight.

September 27. Long zigzag walk amid the old plantations, a few of which are still cultivated in the old way by the same negroes that worked them before the war, and who still occupy their former "quarters." They are now paid seven to ten dollars a month.

The weather is very hot on these sandy, lightly shaded, lowland levels. When very thirsty I discovered a beautiful spring in a sandstone basin overhung with shady bushes and vines, where I enjoyed to the utmost the blessing of pure cold water. Discovered here a fine southern fern, some new grasses, etc. Fancied that I might have been directed here by Providence, while fainting with thirst. It is not often hereabouts that the joys of cool water, cool shade, and rare plants are so delightfully combined.

Witnessed the most gorgeous sunset I ever enjoyed in this bright world of light. The sunny South is indeed sunny. Was directed by a very civil negro to lodgings for the night. Daily bread hereabouts means sweet potatoes and rusty bacon.

September 28. The water oak is abundant on stream banks and in damp hollows. Grasses are becoming tall and cane-like and do not cover the ground with their leaves as at the North. Strange plants are crowding about me now. Scarce a familiar face appears among all the flowers of the day's walk.

September 29. To-day I met a magnificent grass, ten or twelve feet in stature, with a superb panicle of glossy purple flowers. Its leaves, too, are of princely mould and dimensions. Its home is in sunny meadows and along the wet borders of slow streams and swamps. It seems to be fully aware of its high rank, and waves with the grace and solemn majesty of a mountain pine. I wish I could place one of these regal plants among the grass settlements of our Western prairies. Surely every panicle would wave and bow in joyous allegiance and acknowledge their king.

September 30. Between Thomson and Augusta I found many new and beautiful grasses, tall gerardias, liatris, club mosses, etc. Here, too, is the northern limit of the remarkable long-leafed pine, a tree from sixty to seventy feet in height, from twenty to thirty inches in

A SOUTHERN PINE

SPANISH MOSS (*Tillandsia*)

diameter, with leaves ten to fifteen inches long, in dense radiant masses at the ends of the naked branches. The wood is strong, hard, and very resinous. It makes excellent ship spars, bridge timbers, and flooring. Much of it is shipped to the West India Islands, New York, and Galveston.

The seedlings, five or six years old, are very striking objects to one from the North, consisting, as they do, of the straight, leafless stem, surmounted by a crown of deep green leaves, arching and spreading like a palm. Children fancy that they resemble brooms, and use them as such in their picnic play-houses. *Pinus palustris* is most abundant in Georgia and Florida.

The sandy soil here is sparingly seamed with rolled quartz pebbles and clay. Denudation, going on slowly, allows the thorough removal of these clay seams, leaving only the sand. Notwithstanding the sandiness of the soil, much of the surface of the country is covered with standing water, which is easily accounted for by the presence of the above-mentioned impermeable seams.

Traveled to-day more than forty miles without dinner or supper. No family would receive me, so I had to push on to Augusta. Went hungry to bed and awoke with a sore stomach — sore, I suppose, from its walls rubbing on each other without anything to grind. A negro kindly directed me to the best hotel, called, I think, the Planter's. Got a good bed for a dollar.

October 1. Found a cheap breakfast in a market-place; then set off along the Savannah River to Savannah. Splendid grasses and rich, dense, vine-clad forests. Muscadine grapes in cart-loads. Asters and solidagoes becoming scarce. Carices [sedges] quite rare. Leguminous plants abundant. A species of passion flower is common, reaching back into Tennessee. It is here called "apricot vine," has a superb flower, and the most delicious fruit I have ever eaten.

The pomegranate is cultivated here. The fruit is about the size of an orange, has a thick, tough skin, and when opened resembles a many-chambered box full of translucent purple candies.

Toward evening I came to the country of one of the most striking of southern plants, the so-called "Long Moss" or Spanish Moss [Tillandsia], though it is a flowering plant and belongs to the same family as the pineapple [Bromelworts]. The trees hereabouts have all their branches draped with it, producing a remarkable effect.

Here, too, I found an impenetrable cypress swamp. This remarkable tree, called cypress, is a taxodium, grows large and high, and is remarkable for its flat crown. The whole forest seems almost level on the top, as if each tree had grown up against a ceiling, or had been rolled while growing. This taxodium is the only level-topped tree that I have seen. The branches, though spreading, are careful not to pass each other, and stop suddenly on reaching the general level, as if they had grown up against a ceiling.

The groves and thickets of smaller trees are full of blooming evergreen vines. These vines are not arranged in separate groups, or in delicate wreaths, but in bossy walls and heavy, mound-like heaps and banks. Am made to feel that I am now in a strange land. I know hardly any of the plants, but few of the birds, and I am unable to see the country for the solemn, dark, mysterious cypress woods which cover everything.

The winds are full of strange sounds, making one feel far from the people and plants and fruitful fields of home. Night is coming on and I am filled with indescribable loneliness. Felt feverish; bathed in a black, silent stream; nervously watchful for alligators. Obtained lodging in a planter's house among cotton fields. Although the family seemed to be pretty well-off, the only light in the house was bits of pitch-pine wood burned in the fireplace.

October 2. In the low bottom forest of the Savannah River. Very busy with new specimens. Most exquisitely planned wrecks of

Agrostis scabra [Rough Hair Grass]. Pines in glorious array with open, welcoming, approachable plants.

Met a young African with whom I had a long talk. Was amused with his eloquent narrative of coon hunting, alligators, and many superstitions. He showed me a place where a railroad train had run off the track, and assured me that the ghosts of the killed may be seen every dark night.

Had a long walk after sundown. At last was received at the house of Dr. Perkins. Saw Cape Jasmine [*Gardenia florida*] in the garden. Heard long recitals of war happenings, discussion of the slave question, and Northern politics; a thoroughly characteristic Southern family, refined in manners and kind, but immovably prejudiced on everything connected with slavery.

The family table was unlike any I ever saw before. It was circular, and the central part of it revolved. When any one wished to be helped, he placed his plate on the revolving part, which was whirled around to the host, and then whirled back with its new load. Thus every plate was revolved into place, without the assistance of any of the family.

October 3. In "pine barrens" most of the day. Low, level, sandy tracts; the pines wide apart; the sunny spaces between full of beautiful abounding grasses, liatris, long, wandlike solidago, saw palmettos, etc., covering the ground in garden style. Here I sauntered in delightful freedom, meeting none of the catclawed vines, or shrubs, of the alluvial bottoms. Dwarf live-oaks common.

Toward evening I arrived at the home of Mr. Cameron, a wealthy planter, who had large bands of slaves at work in his cotton fields. They still call him "Massa." He tells me that labor costs him less now than it did before the emancipation of the negroes. When I arrived I found him busily engaged in scouring the rust off some cotton-gin saws which had been lying for months at the bottom of his mill-pond to prevent Sherman's "bummers" from destroying them. The most valuable parts of the grist-mill and cotton-press were hidden in the same way. "If Bill Sherman," he said, "should come down now without his army, he would never go back."

When I asked him if he could give me food and lodging for the night he said, "No, no, we have no accommodations for travelers." I said, "But I am traveling as a botanist and either have to find lodgings when night overtakes me or lie outdoors, which I often have had to do in my long walk from Indiana. But you see that the country here is very swampy; if you will at least sell me a piece of bread, and give me a drink at your well, I shall have to look around for a dry spot to lie down on."

Then, asking me a few questions, and narrowly examining me, he said, "Well, it is barely possible that we may find a place for you, and if you will come to the house I will ask my wife." Evidently he was cautious to get his wife's opinion of the kind of creature I was before committing himself to hospitality. He halted me at the door and called out his wife, a fine-looking woman, who also questioned me narrowly as to my object in coming so far down through the South, so soon after the war. She said to her husband that she thought they could, perhaps, give me a place to sleep.

After supper, as we sat by the fire talking on my favorite subject of botany, I described the country I had passed through, its botanical character, etc. Then, evidently, all doubt as to my being a decent man vanished, and they both said that they would n't for anything have turned me away; but I must excuse their caution, for perhaps fewer than one in a hundred, who passed through this unfrequented part of the country, were to be relied upon. "Only a short time ago we entertained a man who was well spoken and well dressed, and he vanished some time during the night with some valuable silverware."

Mr. Cameron told me that when I arrived he tried me for a Mason, and finding that I was not a Mason he wondered still more that I would venture into the country without being able to gain the assistance of brother Masons in these troublous times.

"Young man," he said, after hearing my talks on botany, "I see that your hobby is botany. My hobby is e-lec-tricity. I believe that the time is coming, though we may not live to see it, when that mysterious power or force, used now only for telegraphy, will eventually supply the power for running railroad trains and steamships, for lighting, and, in a word, electricity will do all the work of the world."

Many times since then I have thought of the wonderfully correct vision of this Georgia planter, so far in advance of almost everybody else in the world. Already nearly all that he foresaw has been accomplished, and the use of electricity is being extended more and more every year.

October 4. New plants constantly appearing. All day in dense, wet, dark, mysterious forest of flat-topped taxodiums.

October 5. Saw the stately banana for the

first time, growing luxuriantly in the wayside gardens. At night with a very pleasant, intelligent Savannah family, but as usual was admitted only after I had undergone a severe course of questioning.

October 6. Immense swamps, still more completely fenced and darkened, that are never ruffled with winds or scorched with drought. Many of them seem to be thoroughly aquatic.

October 7. Impenetrable taxodium swamp, seemingly boundless. The silvery skeins of tillandsia becoming longer and more abundant. Passed the night with a very pleasant family of Georgians, after the usual questions and cross questions.

October 8. Found the first woody *compositæ*, a most notable discovery. Took them to be such at a considerable distance. Almost all trees and shrubs are evergreens here with thick polished leaves. *Magnolia grandiflora* becoming common. A magnificent tree in fruit and foliage as well as in flower. Near Savannah I found waste places covered with a dense growth of woody leguminous plants, eight or ten feet high, with pinnate leaves and suspended rattling pods.

Reached Savannah, but find no word from home, and the money that I had ordered to be sent by express from Portage [Wisconsin] by my brother had not yet arrived. Feel dreadfully lonesome and poor. Went to the meanest looking lodging-house that I could find, on account of its cheapness.

CHAPTER IV

CAMPING AMONG THE TOMBS

OCTOBER 9. After going again to the express office and post office, and wandering about the streets, I found a road which led me to the Bonaventure graveyard. If that burying-ground across the Sea of Galilee, mentioned in Scripture, was half as beautiful as Bonaventure, I do not wonder that a man should dwell among the tombs. It is only three or four miles from Savannah, and is reached by a smooth white shell road.

There is but little to be seen on the way in land, water, or sky, that would lead one to hope for the glories of Bonaventure. The ragged desolate fields, on both sides of the road, are overrun with coarse rank weeds, and show scarce a trace of cultivation. But soon all is changed. Rickety log huts, broken fences, and the last patch of weedy rice-stubble are left behind. You come to beds of purple liatris and living wild-wood trees. You hear the song of birds, cross a small stream, and are with Nature in the grand old forest graveyard, so beautiful that almost any sensible person would choose to dwell here with the dead rather than with the lazy, disorderly living.

Part of the grounds was cultivated and planted with live-oak, about a hundred years ago, by a wealthy gentleman who had his country residence here. But much the greater part is undisturbed. Even those spots which are disordered by art, Nature is ever at work to reclaim, and to make them look as if the foot of man had never known them. Only a small plot of ground is occupied with graves and the old mansion is in ruins.

The most conspicuous glory of Bonaventure is its noble avenue of live-oaks. They are the most magnificent planted trees I have ever seen, about fifty feet high and perhaps three or four feet in diameter, with broad spreading leafy heads. The main branches reach out horizontally until they come together over the driveway, embowering it throughout its entire length, while each branch is adorned like a garden with ferns, flowers, grasses, and dwarf palmettos.

But of all the plants of these curious tree-gardens the most striking and characteristic is the so-called Long Moss (*Tillandsia usneoides*). It drapes all the branches from top to bottom, hanging in long silvery-gray skeins, reaching a length of not less than eight or ten feet, and when slowly waving in the wind they produce a solemn funereal effect singularly impressive.

There are also thousands of smaller trees and clustered bushes, covered almost from sight in the glorious brightness of their own light. The place is half surrounded by the salt marshes and islands of the river, their reeds and sedges making a delightful fringe. Many bald eagles roost among the trees along the side of the marsh. Their screams are heard every morning, joined with the noise of crows and the songs of countless warblers, hidden deep in their dwellings of leafy bowers. Large flocks of butter-

IN BONAVENTURE CEMETERY, SAVANNAH

flies, all kinds of happy insects, seem to be in a perfect fever of joy and sportive gladness. The whole place seems like a center of life. The dead do not reign there alone.

Bonaventure to me is one of the most impressive assemblages of animal and plant creatures I ever met. I was fresh from the Western prairies, the garden-like openings of Wisconsin, the beech and maple and oak woods of Indiana and Kentucky, the dark mysterious Savannah cypress forests; but never since I was allowed to walk the woods have I found so impressive a company of trees as the tillandsia-draped oaks of Bonaventure.

I gazed awe-stricken as one new-arrived from another world. Bonaventure is called a graveyard, a town of the dead, but the few graves are powerless in such a depth of life. The rippling of living waters, the song of birds, the joyous confidence of flowers, the calm, undisturbable grandeur of the oaks, mark this place of graves as one of the Lord's most favored abodes of life and light.

On no subject are our ideas more warped and pitiable than on death. Instead of the sympathy, the friendly union, of life and death so apparent in Nature, we are taught that death is an accident, a deplorable punishment for the oldest sin, the arch-enemy of life, etc. Town children, especially, are steeped in this death orthodoxy, for the natural beauties of death are seldom seen or taught in towns.

Of death among our own species, to say nothing of the thousand styles and modes of murder, our best memories, even among happy deaths, yield groans and tears, mingled with morbid exultation; burial companies, black in cloth and countenance; and, last of all, a black box burial in an ill-omened place, haunted by imaginary glooms and ghosts of every degree. Thus death becomes fearful, and the most notable and incredible thing heard around a death-bed is, "I fear not to die."

But let children walk with Nature, let them see the beautiful blendings and communions of death and life, their joyous inseparable unity, as taught in woods and meadows, plains and mountains and streams of our blessed star, and they will learn that death is stingless indeed, and as beautiful as life, and that the grave has no victory, for it never fights. All is divine harmony.

Most of the few graves of Bonaventure are planted with flowers. There is generally a magnolia at the head, near the strictly erect marble, a rose-bush or two at the foot, and some violets and showy exotics along the sides or on the tops. All is enclosed by a black iron railing, composed of rigid bars that might have been spears or bludgeons from a battlefield in Pandemonium.

It is interesting to observe how assiduously Nature seeks to remedy these labored art blunders. She corrodes the iron and marble, and gradually levels the hill which is always heaped up, as if a sufficiently heavy quantity of clods could not be laid on the dead. Arching grasses come one by one; seeds come flying on downy wings, silent as fate, to give life's dearest beauty for the ashes of art; and strong evergreen arms laden with ferns and tillandsia drapery are spread over all — Life at work everywhere, obliterating all memory of the confusion of man.

In Georgia many graves are covered with a common shingle roof, supported on four posts as the cover of a well, as if rain and sunshine were not regarded as blessings. Perhaps, in this hot and insalubrious climate, moisture and sunheat are considered necessary evils to which they do not wish to expose their dead.

The money package that I was expecting did not arrive until the following week. After stopping the first night at the cheap, disreputable-looking hotel, I had only about a dollar and a half left in my purse, and so was compelled to camp out to make it last in buying only bread. I went out of the noisy town to seek a sleeping-place that was not marshy. After gaining the outskirts of the town toward the sea, I found some low sand dunes, yellow with flowering solidagoes.

I wandered wearily from dune to dune sink-ing ankle-deep in the sand, searching for a place to sleep beneath the tall flowers, free from insects and snakes, and above all from my fellow man. But idle negroes were prowling about everywhere, and I was afraid. The wind had strange sounds, waving the heavy panicles over my head, and I feared sickness from malaria so prevalent here, when I suddenly thought of the graveyard.

"There," thought I, "is an ideal place for a penniless wanderer. There no superstitious prowling mischief maker dares venture for fear of haunting ghosts, while for me there will be God's rest and peace. And then, if I am to be exposed to unhealthy vapors, I shall have capital compensation in seeing those grand oaks in the moonlight, with all the impressive and nameless influences of this lonely beautiful place."

By this time it was near sunset, and I hastened across the common to the road and set off for Bonaventure, delighted with my choice, and almost glad to find that necessity had furnished me with so good an excuse for doing what I knew my mother would censure; for she made me promise I would not lie out of doors if I could possibly avoid it. The sun was set ere I was past the negroes' huts and rice fields, and I arrived near the graves in the silent hour of the gloaming.

I was very thirsty after walking so long in the muggy heat, a distance of three or four miles from the city, to get to this graveyard. A dull, sluggish, coffee-colored stream flows under the road just outside the graveyard garden park, from which I managed to get a drink after breaking a way down to the water through a dense fringe of bushes, daring the snakes and alligators in the dark. Thus refreshed I entered the weird and beautiful abode of the dead.

All the avenue where I walked was in shadow, but an exposed tombstone frequently shone out in startling whiteness on either hand, and thickets of sparkleberry bushes gleamed like heaps of crystals. Not a breath of air moved the gray moss, and the great black arms of the trees met overhead and covered the avenue. But the canopy was fissured by many a netted seam and leafy-edged opening, through which the moonlight sifted in auroral rays, broidering the blackness in silvery light. Though tired, I sauntered a while enchanted, then lay down under one of the great oaks. I found a little mound that served for a pillow, placed my plant press and bag beside me and rested fairly well, though somewhat disturbed by large prickly-footed beetles creeping across my hands and face, and by a lot of hungry stinging mosquitoes.

When I awoke, the sun was up and all Nature was rejoicing. Some birds had discovered me as an intruder, and were making a great ado in interesting language and gestures. I heard the screaming of the bald eagles, and of some strange waders in the rushes. I heard the hum of Savannah with the long jarring hallos of negroes far away. On rising I found that my head had been resting on a grave, and though my sleep had not been quite so sound as that of the person below, I arose refreshed, and looking about me, the morning sunbeams pouring through the oaks and gardens dripping with dew, the beauty displayed was so glorious and exhilarating that hunger and care seemed only a dream.

Eating a breakfast cracker or two and watching for a few hours the beautiful light, birds, squirrels, and insects, I returned to Savannah, to find that my money package had not yet arrived. I then decided to go early to the graveyard and make a nest with a roof to keep off the dew, as there was no way of finding out how long I might have to stay. I chose a hidden spot in a dense thicket of sparkleberry bushes, near the right bank of the Savannah River, where the bald eagles and a multitude of singing birds roosted. It was so well hidden that I had to carefully fix its compass bearing in my mind from a mark I made on the side of the main avenue, that I might be able to find it at bedtime.

I used four of the bushes as corner posts for

my little hut, which was about four or five feet long by about three or four in width, tied little branches across from forks in the bushes to support a roof of rushes, and spread a thick mattress of Long Moss over the floor for a bed. My whole establishment was on so small a scale that I could have taken up, not only my bed, but my whole house, and walked. There I lay that night, eating a few crackers.

Next day I returned to the town and was disappointed as usual in obtaining money. So after spending the day looking at the plants in the gardens of the fine residences and town squares, I returned to my graveyard home. That I might not be observed and suspected of hiding, as if I had committed a crime, I always went home after dark, and one night, as I lay down in my moss nest, I felt some cold-blooded creature in it; whether a snake or simply a frog or toad I do not know, but instinctively, instead of drawing back my hand, I grasped the poor creature and threw it over the tops of the bushes. That was the only significant disturbance or fright that I got.

In the morning everything seemed divine. Only squirrels, sunbeams, and birds came about me. I was awakened every morning by these little singers after they discovered my nest. Instead of serenely singing their morning songs they at first came within two or three feet of the hut, and, looking in at me through the leaves, chattered and scolded in half-angry, half-wondering tones. The crowd constantly increased, attracted by the disturbance. Thus I began to get acquainted with my bird neighbors in this blessed wilderness, and after they learned that I meant them no ill they scolded less and sang more.

After five days of this graveyard life I saw that even with living on three or four cents a day my last twenty-five cents would soon be spent, and after trying again and again unsuccessfully to find some employment began to think that I must strike farther out into the country, but still within reach of town, until

I came to some grain or rice field that had not yet been harvested, trusting that I could live indefinitely on toasted or raw corn, or rice.

By this time I was becoming faint, and in making the journey to the town was alarmed to find myself growing staggery and giddy. The ground ahead seemed to be rising up in front of me, and the little streams in the ditches on the sides of the road seemed to be flowing up hill. Then I realized that I was becoming dangerously hungry and became more than ever anxious to receive that money package.

To my delight this fifth or sixth morning, when I inquired if the money package had come, the clerk replied that it had, but that he could not deliver it without my being identified. I said, "Well, here! read my brother's letter," handing it to him. "It states the amount in the package, where it came from, the day it was put into the office at Portage City, and I should think that would be enough." He said, "No, that is not enough. How do I know that this letter is yours? You may have stolen it. How do I know that you are John Muir?"

I said, "Well, don't you see that this letter indicates that I am a botanist? For in it my brother says, 'I hope you are having a good time and finding many new plants.' Now, you say that I might have stolen this letter from John Muir, and in that way have become aware of there being a money package to arrive from Portage for him. But the letter proves that John Muir must be a botanist, and though, as you say, his letter might have been stolen, it would hardly be likely that the robber would be able to steal John Muir's knowledge of botany. Now I suppose, of course, that you have been to school and know something of botany. Examine me and see if I know anything about it."

At this he laughed good-naturedly, evidently feeling the force of my argument, and, perhaps, pitying me on account of looking pale and hungry, he turned and rapped at the door of a private office — probably the Manager's — called him out and said, "Mr. So and So, here is a man who has inquired every day for the last week or so for a money package from Portage, Wisconsin. He is a stranger in the city with no one to identify him. He states correctly the amount and the name of the sender. He has shown me a letter which indicates that Mr. Muir is a botanist, and that although a traveling companion may have stolen Mr. Muir's letter, he could not have stolen his botany, and requests us to examine him."

The head official smiled, took a good stare into my face, waved his hand, and said, "Let him have it." Gladly I pocketed my money, and had not gone along the street more than a few rods before I met a very large negro woman with a tray of gingerbread, in which I immediately invested some of my new wealth, and walked rejoicingly, munching along the street, making no attempt to conceal the pleasure I had in eating. Then, still hunting for more food, I found a sort of eating-place in a market and had a large regular meal on top of the gingerbread! Thus my "marching through Georgia" terminated handsomely in a jubilee of bread.

CHAPTER V

THROUGH FLORIDA SWAMPS AND FORESTS

OF the people of the States that I have now passed, I best like the Georgians. They have charming manners, and their dwellings are mostly larger and better than those of adjacent States. However costly or ornamental their homes or their manners, they do not, like those of the New Englander, appear as the fruits of intense and painful sacrifice and training, but are entirely divested of artificial weights and measures, and seem to pervade and twine about their characters as spontaneous growths with the durability and charm of living nature.

In particular, Georgians, even the commonest, have a most charmingly cordial way of saying to strangers, as they proceed on their journey, "I wish you well, sir." The negroes of Georgia, too, are extremely mannerly and polite, and appear always to be delighted to find opportunity for obliging anybody.

Athens contains many beautiful residences. I never before saw so much about a home that was so evidently done for beauty only, although this is by no means a universal characteristic of Georgian homes. Nearly all well-to-do farmers' families in Georgia and Tennessee spin and weave their own cloth. This work is almost all done by the mothers and daughters and consumes much of their time.

The traces of war are not only apparent on the broken fields, burnt fences, mills, and woods ruthlessly slaughtered, but also on the countenances of the people. A few years after a forest has been burned another generation of bright and happy trees arises, in purest, freshest vigor; only the old trees, wholly or half dead, bear marks of the calamity. So with the people of this war-field. Happy, unscarred, and unclouded youth is growing up around the aged, half-consumed, and fallen parents, who bear in sad measure the ineffaceable marks of the farthest-reaching and most infernal of all civilized calamities.

Florida Swamps and Forests

Since the commencement of my floral pilgrimage I have seen much that is not only new, but altogether unallied, unacquainted with the plants of my former life. I have seen magnolias, tupelo, live-oak, Kentucky oak, tillandsia, long-leafed pine, palmetto, schrankia, and whole forests of strange trees and vine-tied thickets of blooming shrubs; whole meadowfuls of magnificent bamboo and lakefuls of lilies, all new to me; yet I still press eagerly on to Florida as the special home of the tropical plants I am looking for, and I feel sure I shall not be disappointed.

The same day on which the money arrived I took passage on the steamship Sylvan Shore for Fernandina, Florida. The daylight part of this sail along the coast of Florida was full of novelty, and by association awakened memories of my Scottish days at Dunbar on the Firth of Forth.

On board I had civilized conversation with a Southern planter on topics that are found floating in the mind of every white man down here who has a single thought. I also met a brother Scotchman, who was especially interesting and had some ideas outside of Southern politics. Altogether my half-day and night on board the steamer were pleasant, and carried me past a very sickly, entangled, overflowed, and unwalkable piece of forest.

It is pretty well known that a short geological time ago the ocean covered the sandy level margin, extending from the foot of the Alleghanies to the present coast-line, and in receding left many basins for lakes and swamps. The land is still encroaching on the sea, and it does so not evenly, in a regular line, but in fringing lagoons and inlets and dotlike coral islands.

It is on the coast strip of isles and peninsulas that sea-island cotton is grown. Some of these small islands are afloat, anchored only by the roots of mangroves and rushes. For a few hours our steamer sailed in the open sea, exposed to its waves, but most of the time she threaded her way among the lagoons, the home of alligators and countless ducks and waders.

October 15. To-day, at last, I reached Florida, the so-called "Land of Flowers," that I had so long waited for, wondering if after all my longings and prayers would be in vain, and I should die without a glimpse of the flowery Canaan. But here it is, at the distance of a few yards! — a flat, watery, reedy coast, with clumps of mangrove and forests of moss-dressed, strange trees appearing low in the distance. The steamer finds her way among the reedy islands like a duck, and I step on a rickety wharf. A few steps more take me to a rickety town, Fernandina. I discover a baker, buy some bread, and without asking a single question, make for the shady, gloomy groves.

In visiting Florida in dreams, of either day or night, I always came suddenly on a close forest of trees, every one in flower, and bent down and entangled to network by luxuriant, bright-blooming vines, and over all a flood of bright sunlight. But such was not the gate by which I entered the promised land. Salt marshes, belonging more to the sea than to the land; with groves here and there, green and unflowered, sunk to the shoulders in sedges and rushes; with trees farther back, ill defined in their boundary, and instead of rising in hilly waves and swellings, stretching inland in low water-like levels.

We were all discharged by the captain of the steamer without breakfast, and, after meeting and examining the new plants that crowded about me, I threw down my press and little bag beneath a thicket, where there was a dry spot on some broken heaps of grass and roots, something like a deserted muskrat house, and applied myself to my bread breakfast. Everything in earth and sky had an impression of strangeness; not a mark of friendly recognition, not a breath, not a spirit whisper of sympathy came from anything about me, and of course I was lonely. I lay on my elbow eating my bread, gazing, and listening to the profound strangeness.

While thus engaged I was startled from these gatherings of melancholy by a rustling sound in the rushes behind me. Had my mind been in health, and my body not starved, I should only have turned calmly to the noise. But in this half-starved, unfriended condition I could have no healthy thought, and I at once believed that the sound came from an alligator. I fancied I could feel the stroke of his long notched tail, and could see his big jaws and rows of teeth, closing with a springy snap on me, as I had seen in pictures.

Well, I don't know the exact measure of my fright either in time or pain, but when I did come to a knowledge of the truth, my man-eating alligator became a tall white crane, handsome as a minister from spirit land — "only that." I was ashamed and tried to excuse myself on account of Bonaventure anxiety and hunger.

Florida is so watery and vine-tied that pathless wanderings are not easily possible in any direction. I started to cross the State by a gap hewn for the locomotive, walking sometimes between the rails, stepping from tie to tie, or walking on the strip of sand at the sides, gazing into the mysterious forest, Nature's own. It is impossible to write the dimmest picture of plant grandeur so redundant, unfathomable.

Short was the measure of my walk to-day. A new, canelike grass, or big lily, or gorgeous flower belonging to tree or vine, would catch my attention, and I would throw down my bag and press and splash through the coffee-brown water for specimens. Frequently I sank deeper and deeper until compelled to turn back and make the attempt in another and still another place. Oftentimes I was tangled in a labyrinth of armed vines like a fly in a spider-web. At all times, whether wading or climbing a tree for specimens of fruit, I was overwhelmed with the vastness and unapproachableness of the great guarded sea of sunny plants.

Magnolia grandiflora I had seen in Georgia; but its home, its better land, is here. Its large dark-green leaves, glossy bright above

BY THE ST. JOHN'S RIVER, IN EASTERN FLORIDA

and rusty brown beneath, gleam and mirror the sunbeams most gloriously among countless flower-heaps of the climbing, smothering vines. It is bright also in fruit and more tropical in form and expression than the orange. It speaks itself a prince among its fellows.

Occasionally, I came to a little strip of open sand, planted with pine (*Pinus palustris* or *Cubensis*). Even these spots were mostly wet, though lighted with free sunshine, and adorned with purple liatris, and orange-colored *Osmunda cinnamomea*. But the grandest discovery of this great wild day was the palmetto.

I was meeting so many strange plants that I was much excited, making many stops to get specimens. But I could not force my way far through the swampy forest, although so tempting and full of promise. Regardless of water snakes or insects, I endeavored repeatedly to force a way through the tough vine-tangles, but seldom succeeded in getting farther than a few hundred yards.

It was while feeling sad to think that I was

only walking on the edge of the vast wood, that I caught sight of the first palmetto in a grassy place, standing almost alone. A few magnolias were near it, and bald cypresses, but it was not shaded by them. They tell us that plants are perishable, soulless creatures, that only man is immortal, etc.; but this, I think, is something that we know very nearly nothing about. Anyhow, this palm was indescribably impressive and told me grander things than I ever got from human priest.

This vegetable has a plain gray shaft, round as a broom-handle, and a crown of varnished channeled leaves. It is a plainer plant than the humblest of Wisconsin oaks; but, whether rocking and rustling in the wind or poised thoughtful and calm in the sunshine, it has a power of expression not excelled by any plant high or low that I have met in my whole walk thus far.

This, my first specimen, was not very tall, only about twenty-five feet high, with fifteen or twenty leaves, arching equally and evenly all around. Each leaf was about ten feet in length, the blade four feet, the stalk six. The leaves are channeled like half-open clams and are highly polished, so that they reflect the sunlight like glass. The undeveloped leaves on the top stand erect, closely folded, all together forming an oval crown over which the tropic light is poured and reflected from its slanting mirrors in sparks and splinters and long-rayed stars.

I am now in the hot gardens of the sun, where the palm meets the pine, longed and prayed for and often visited in dreams, and, though lonely to-night amid this multitude of strangers, strange plants, strange winds blowing gently, whispering, cooing, in a language I never learned, and strange birds also, everything solid or spiritual full of influences that I never before felt, yet I thank the Lord with all my heart for his goodness in granting me admission to this magnificent realm.

October 16. Last evening when I was in the trackless woods, the great mysterious night becoming more mysterious in the thickening darkness, I gave up hope of finding food or a house bed, and searched only for a dry spot on which to sleep safely hidden from wild, runaway negroes. I walked rapidly for hours in the wet, level woods, but not a foot of dry ground could I find. Hollow-voiced owls were calling without intermission. All manner of night sounds came from strange insects and beasts, one by one, or crowded together. All had a home but I. Jacob on the dry plains of Padan-aram, with a stone pillow, must have been comparatively happy.

When I came to an open place where pines grew, it was about ten o'clock, and I thought that now at last I would find dry ground. But even the sandy barren was wet, and I had to grope in the dark a long time, feeling the ground with my hands when my feet ceased to plash, before I at last discovered a little hillock dry enough to lie down on. I ate a piece of bread that I fortunately had in my bag, drank some of the brown water about my precious hillock, and lay down. The noisiest of the unseen witnesses around me were the owls, who pronounced their gloomy speeches with profound emphasis, but did not prevent the coming of sleep to heal weariness.

In the morning I was cold and wet with dew, and I set out breakfastless. Flowers and beauty I had in abundance, but no bread. A serious matter is this bread which perishes, and, could it be dispensed with, I doubt if civilization would ever see me again. I walked briskly, watching for a house, as well as the grand assemblies of novel plants.

Near the middle of the forenoon I came to a shanty where a party of loggers were getting out long pines for ship spars. They were the wildest of all the white savages I have met. The long-haired ex-guerrillas of the mountains of Tennessee and North Carolina are uncivilized fellows; but for downright barbarism these Florida loggers excel. Nevertheless, they gave me a portion of their yellow pork and hominy without either apparent hospitality or a grudge, and I was glad to escape to the forest again.

A few hours later I dined with three men and

three dogs. I was viciously attacked by the latter, who undertook to undress me with their teeth. I was nearly dragged down backward, but escaped unbitten. Liver pie, mixed with sweet potatoes and fat duff, was set before me, and after I had finished a moderate portion, one of the men, turning to his companion, remarked: "Wall, I guess that man quit eatin' 'cause he had nothin' more to eat. I'll get him more potato."

Arrived at a place on the margin of a stagnant pool where an alligator had been rolling and sunning himself. "See," said a man who lived here, "see, what a track that is! He must have been a mighty big fellow. Alligators wallow like hogs and like to lie in the sun. I'd like a shot at that fellow." Here followed a long recital of bloody combats with the scaly enemy, in many of which he had, of course, taken an important part. Alligators are said to be extremely fond of negroes and dogs, and naturally the dogs and negroes are afraid of them.

Another man that I met to-day pointed to a

shallow, grassy pond before his door. "There," said he, "I once had a tough fight with an alligator. He caught my dog. I heard him howling, and as he was one of my best hunters I tried hard to save him. The water was only about knee-deep and I ran up to the alligator. It was only a small one about four feet long, and was having trouble in its efforts to drown the dog in the shallow water. I scared him and made him let go his hold, but before the poor crippled dog could reach the shore, he was caught again, and when I went at the alligator with a knife, it seized my arm. If it had been a little stronger it might have eaten me instead of my dog."

I never in all my travels saw more than one, though they are said to be abundant in most of the swamps, and frequently attain a length of nine or ten feet. It is reported, also, that they are very savage, oftentimes attacking men in boats. These independent inhabitants of the sluggish waters of this low coast cannot be called the friends of man, though I heard of one big fellow that was caught young and was partially civilized and made to work in harness.

Many good people believe that alligators were created by the Devil, thus accounting for their all-consuming appetite and ugliness. But doubtless these creatures are happy and fill the place assigned them by the great Creator of us all. Fierce and cruel they appear to us, but beautiful in the eyes of God. They, also, are his children, for He hears their cries, cares for them tenderly, and provides their daily bread.

The antipathies existing in the Lord's great animal family must be wisely planned, like balanced repulsion and attraction in the mineral kingdom. How narrow we selfish, conceited creatures are in our sympathies! how blind to the rights of all the rest of creation! With what dismal irreverence we speak of our fellow mortals! Though alligators, snakes, etc., naturally repel us, they are not mysterious evils. They dwell happily in these flowery wilds, are part of God's family, unfallen, undepraved, and cared for with the same species of tenderness and love as is bestowed on angels in heaven or saints on earth.

I think that most of the antipathies which haunt and terrify us are morbid productions of ignorance and weakness. I have better thoughts of those alligators now that I have seen them at home. Honorable representatives of the great saurians of an older creation, may you long enjoy your lilies and rushes, and be blessed now and then with a mouthful of terror-stricken man by way of dainty!

Found a beautiful lycopodium to-day, and many grasses in the dry sunlit places called "barrens," "hummocks," "savannas," etc. Ferns also are abundant. What a flood of heat and light is daily poured out on these beautiful openings and intertangled woods! "The land of the sunny South," we say, but no part of our diversified country is more shaded and covered from sunshine. Many a sunny sheet of plain and prairie break the continuity of the forests of the North and West, and the forests themselves are mostly lighted also, pierced with direct ray lances, or [the sunlight] passing to the earth and the lowly plants in filtered softness through translucent leaves. But in the dense Florida forests sunlight cannot enter. It falls on the evergreen roof and rebounds in long silvery lances and flashy spray. In many places there is not light sufficient to feed a single green leaf on these dark forest floors. All that the eye can reach is just a maze of tree stems and crooked leafless vine strings. All the flowers, all the verdure, all the glory is up in the light.

The streams of Florida are still young, and in many places are untraceable. I expected to find these streams a little discolored from the vegetable matter that I knew they must contain, and I was sure that in so flat a country I should not find any considerable falls or long rapids. The streams of upper Georgia are almost unapproachable in some places on account of luxuriant bordering vines, but the banks are nevertheless high and well defined. Florida streams are not yet possessed of banks and braes and definite channels. Their waters in deep places are black as ink, perfectly opaque, and glossy on the surface as if varnished. It often is difficult to ascertain which way they are flowing or creeping, so slowly and so widely do they circulate through the tree-tangles and swamps of the woods. The flowers here are strangers to me, but not more so than the rivers and lakes. Most streams appear to travel through a country with thoughts and plans for something beyond. But those of Florida are at home, do not appear to be traveling at all, and seem to know nothing of the sea.

October 17. Found a small, silvery-leafed magnolia, a bush ten feet high. Passed through a good many miles of open level pine barrens, as bounteously lighted as the "openings" of Wisconsin. The pines are rather small, are planted sparsely and pretty evenly on these sandy flats not long risen from the sea. Scarcely a specimen of any other tree is to be found associated with the pine. But there are some thickets of the little saw palmettos and a magnificent assemblage of tall grasses, their splendid panicles waving grandly in the warm wind, and making low tuneful changes in the glistening light that is flashed from their bent stems.

Not a pine, not a palm, in all this garden excels these stately grass plants in beauty of wind-waving gestures. Here are panicles that are one mass of refined purple; others that have flowers as yellow as ripe oranges, and stems polished and shining like steel wire. Some of the species are grouped in groves and thickets like trees, while others may be seen waving without any companions in sight. Some of them have wide-branching panicles like Kentucky oaks, others with a few tassels of spikelets drooping from a tall, leafless stem. But all of them are beautiful beyond the reach of language. I rejoice that God has "so clothed the grass of the field." How strangely we are blinded to beauty and color, form and motion, by comparative size! For example, we measure grasses by our own stature and by the height and bulkiness of trees. But what is the size of the greatest man, or the tallest tree that ever overtopped a grass! Compared with other things in God's creation the difference is nothing. We all are only microscopic animalcula.

October 18. Am walking on land that is almost dry. The dead levels are interrupted here and there by sandy waves a few feet in height. It is said that not a point in all Florida is more than three hundred feet above sea-level — a country where but little grading is required for roads, but much bridging, and boring of many tunnels through forests.

Before reaching this open ground, in a lonely, swampy place in the woods, I met a large, muscular, brawny young negro, who eyed me with glaring, wistful curiosity. I was very thirsty at the time, and inquired of the man if there were any houses or springs near by where I could get a drink. "Oh, yes," he replied, still eagerly searching me with his wild eyes. Then he inquired where I came from, where I was going, and what brought me to such a wild country, where I was liable to be robbed, and perhaps killed.

"Oh, I am not afraid of any one robbing me," I said, "for I don't carry anything worth stealing." "Yes," said he, "but you can't travel without money." I started to walk on, but he blocked my way. Then I noticed that he was trembling, and it flashed upon me all at once that he was thinking of knocking me down in order to rob me. After glaring at my pockets as if searching for weapons, he stammered in a quavering voice, "Do you carry shooting-irons?" His motives, which I ought to have noted sooner, now were apparent to me. Though I had no pistol, I instinctively threw my hand back to my pistol pocket and, with my eyes fixed on his, I marched up close to him and said, "I allow people to find out if I am armed or not." Then he quailed, stepped aside, and allowed me to pass, for fear of being shot. This was evidently a narrow escape.

A few miles farther on I came to a cotton-field, to patches of sugar cane carefully fenced, and some respectable-looking houses with gardens. These little fenced fields look as if they were intended to be for plants what cages are for birds. Discovered a large, treelike cactus in a dooryard; a small species was abundant on the sand-hillocks. Reached Gainesville late in the night.

When within three or four miles of the town I noticed a light off in the pine woods. As I was very thirsty, I thought I would venture toward it with the hope of obtaining water. In creeping cautiously and noiselessly through the grass to discover whether or no it was a camp of robber negroes, I came suddenly in full view of the best-lighted and most primitive of all the domestic establishments I have yet seen in town or grove. There was, first of all, a big, glowing log fire, illuminating the overleaning bushes and trees, bringing out leaf and spray with more than noonday distinctness, and making still darker the surrounding wood. In the center of this globe of light sat two negroes. I could see their ivory gleaming from the great lips, and their smooth cheeks flashing off light as if made of glass. Seen anywhere but in the South, the glossy pair would have been taken for twin devils, but here it was only a negro and his wife at their supper.

I ventured forward to the radiant presence of the black pair, and, after being stared at with that desperate fixedness which is said to subdue the lion, I was handed water in a gourd from somewhere out of the darkness. I was standing for a moment beside the big fire, looking at the unsurpassable simplicity of the establishment, and asking questions about the road to Gainesville, when my attention was called to a black lump of something lying in the ashes of the fire. It seemed to be made of rubber; but ere I had time for much speculation, the woman bent wooingly over the black object and said with motherly kindness, "Come, honey, eat yo' hominy."

At the sound of "hominy" the rubber gave strong manifestations of vitality and proved to be a burly little negro boy, rising from the earth naked as to the earth he came. Had he emerged from the black muck of a marsh, we might easily have believed that the Lord had manufactured him like Adam direct from the earth.

Surely, thought I, as I started for Gainesville, surely I am now coming to the tropics, where the inhabitants wear nothing but their own skins. This fashion is sufficiently simple, — "no troublesome disguises," as Milton calls clothing, — but it certainly is not quite in harmony with Nature. Birds make nests and nearly all beasts make some kind of bed for their young; but these negroes allow their younglings to lie nestless and naked in the dirt.

Gainesville is rather attractive — an oasis in the desert, compared with other villages. Its gets its life from the few plantations located about it on dry ground that rises islandlike a few feet above the swamps. Obtained food and lodging at a sort of tavern.

October 19. Dry land nearly all day. Encountered limestone, flint, coral, shells, etc. Passed several thrifty cotton plantations with comfortable residences, contrasting sharply with the squalid hovels of my first days in Florida. Found a single specimen of a handsome little plant, which at once, in some mysterious way, brought to mind a young friend in Indiana. How wonderfully our thoughts and impressions are stored! There is that in the glance of a flower which may at times control the greatest of creation's braggart lords.

The magnolia is much more abundant here. It forms groves and almost exclusively forests the edges of ponds and the banks of streams. The easy, dignified simplicity of this noble tree, its plain leaf endowed with superb richness of color and form, its open branches festooned with graceful vines and tillandsia, its showy crimson fruit, and its magnificent fragrant white flowers make *Magnolia grandiflora* the most lovable of Florida trees.

Discovered a great many beautiful polygonums, petalostemons, and yellow leguminous vines. Passed over fine sunny areas of the long-leafed and Cuban pines, which were everywhere accompanied by fine grasses and solidagoes. Wild orange groves are said to be rather common here, but I have seen only limes growing wild in the woods.

Came to a hut about noon, and, being weary and hungry, asked if I could have dinner. After serious consultation I was told to wait, that dinner would soon be ready. I saw only the man and his wife. If they had children, they may have been hidden in the weeds on account of nakedness. Both were suffering from malarial fever, and were very dirty. But they did not appear to have any realizing sense of discomfort from either the one or the other of these misfortunes. The dirt which encircled the countenances of these people did not, like the common dirt of the North, stick on the skin in bold union like plaster or paint, but appeared to stand out a little on contact like a hazy, misty, half-aerial mud envelope, the most diseased and incurable dirt that I ever saw, evidently desperately chronic and hereditary.

It seems impossible that children from such

parents could ever be clean. Dirt and disease are dreadful enough when separate, but combined are inconceivably horrible. The neat cottage with a fragrant circumference of thyme and honeysuckle is almost unknown here. I have seen dirt on garments regularly stratified, the various strata no doubt indicating different periods of life. Some of them, perhaps, were annual layers, furnishing, like those of trees, a means of determining the age. Man and other civilized animals are the only creatures that ever become dirty.

Slept in the barrens at the side of a log. Suffered from cold and was drenched with dew. What a comfort a companion would be in the dark loneliness of such nights! Did not dare to make a fire for fear of discovery by robber negroes, who, I was warned, would kill a man for a dollar or two. Had a long walk after nightfall, hoping to discover a house. Became very thirsty and often was compelled to drink from slimy pools groped for in the grass, with the fear of alligators before my eyes.

October 20. Swamp very dense during this day's journey. Almost one continuous sheet of water covered with aquatic trees and vines. No stream that I crossed to-day appeared to have the least idea where it was going. Saw an alligator plash into the sedgy brown water by the roadside from an old log.

Arrived at night at the house of Captain Simmons, one of the very few scholarly, intelligent men that I have met in Florida. He had been an officer in the Confederate army in the war and was, of course, prejudiced against the North, but polite and kind to me, nevertheless. Our conversation, as we sat by the light of the fire, was on the one great question, slavery and its concomitants. I managed, however, to switch off to something more congenial occasionally — the birds of the neighborhood, the animals, the climate, and what spring, summer, and winter are like in these parts.

About the climate, I could not get much information, as he had always lived in the South and, of course, saw nothing extraordinary in weather to which he had always been accustomed. But in speaking of animals, he at once became enthusiastic and told many stories of hairbreadth escapes, in the woods about his house, from bears, hungry alligators, wounded deer, etc. "And now," said he, forgetting in his kindness that I was from the hated North, "you must stay with me a few days. Deer are abundant. I will lend you a rifle and we'll go hunting. I hunt whenever I wish venison, and I can get it about as easily from the woods near by as a shepherd can get mutton out of his flock. And perhaps we will see a bear, for they are far from scarce here, and there are some big gray wolves, too."

I expressed a wish to see some large alligators. "Oh, well," said he, "I can take you where you will see plenty of those fellows, but they are not much to look at. I once got a good look at an alligator that was lying at the bottom of still, transparent water, and I think that his eyes were the most impressively cold and cruel of any animal I have seen. Many alligators go out to sea among the keys. These sea alligators are the largest and most ferocious, and sometimes attack people by trying to strike them with their tails when they are out fishing in boats.

"Another thing I wish you to see," he continued, "is a palmetto grove on a rich hummock a few miles from here. The grove is about seven miles in length by three in breadth. The ground is covered with long grass, uninterrupted with bushes or other trees. It is the finest grove of palmettos I have ever seen and I have oftentimes thought that it would make a fine subject for an artist."

I concluded to stop — more to see this wonderful palmetto hummock than to hunt. Besides, I was weary and the prospect of getting a little rest was a tempting consideration after so many restless nights and long, hard walks by day.

October 21. Having outlived the sanguinary hunters' tales of my loquacious host, and breakfasted sumptuously on fresh venison and "caller" fish from the sea, I set out for the grand palm grove. I had seen these dazzling sun-children in every day of my walk through Florida, but they were usually standing solitary, or in groups of three or four; but to-day I was to see them by the mile. The captain led me a short distance through his corn field and showed me a trail which would conduct me to the palmy hummock. He pointed out the general direction, which I noted upon my compass.

"Now," said he, "at the other side of my farthest field you will come to a jungle of cat-briers, but will be able to pass them if you manage to keep the trail. You will find that the way is not by any means well marked, for in passing through a broad swamp, the trail makes a good many abrupt turns to avoid deep water, fallen trees, or impenetrable thickets. You will have to wade a good deal, and in passing the water-covered places you will have to watch for the point where the trail comes out on the opposite side."

I made my way through the briers, which in strength and ferocity equaled those of Tennessee, followed the path through all of its dim waverings, waded the many opposing pools, and, emerging suddenly from the leafy darkness of the swamp forest, at last stood free and unshaded on the border of the sun-drenched palm garden. It was a level area of grasses and sedges, smooth as a prairie, well starred with flowers, and bounded like a clearing by a wall of vine-laden trees.

The palms had full possession and appeared to enjoy their sunny home. There was no jostling, no apparent effort to outgrow each other. Abundance of sunlight was there for every crown, and plenty to fall between. I walked enchanted in their midst. What a landscape! Only palms as far as the eye could reach! Smooth pillars rising from the grass, each capped with a sphere of leaves, shining in the sun as bright as a star. The silence and calm were as deep as ever I found in the dark, solemn pine woods of Canada, and that contentment which is an attribute of the best of God's plant people was as impressively felt in this alligator wilderness as in the homes of the happy, healthy people of the North.

The admirable Linnæus calls palms "the princes of the vegetable world." I know that there is grandeur and nobility in their character, and that there are palms nobler far than these. But in rank they appear to me to stand below both the oak and the pine. The motions of the palms, their gestures, are not very graceful. They appear to best advantage when perfectly motionless in the noontide calm and intensity of light. But they rustle and rock in the evening wind. I have seen grasses waving with far more dignity. And when our northern pines are waving and bowing in sign of worship with the winter storm-winds, where is the prince of palms that could have the conscience to demand their homage!

Members of this palm congregation were of all sizes with respect to their stems; but their glorious crowns were all alike. In develop-

A FLORIDA PALMETTO HUMMOCK, OR "HAMMOCK"

ment there is only the terminal bud to consider. The young palm of this species emerges from the ground in full strength, one cluster of leaves arched every way, making a sphere about ten or twelve feet in diameter. The outside lower leaves gradually become yellow, wither, and break off, the petiole snapping squarely across, a few inches from the stem. New leaves develop with wonderful rapidity. They stand erect at first, but gradually arch outward as they expand their blades and lengthen their petioles.

New leaves arise constantly from the center of the grand bud, while old ones break away from the outside. The splendid crowns are thus kept about the same size, perhaps a little larger than in youth while they are yet on the ground. As the development of the central axis goes on, the crown is gradually raised on a stem of about six to twelve inches in diameter. This stem is of equal thickness at the top and at the bottom and when young is roughened with the broken petioles. But these petiole-stumps fall off and disappear as they become old, and the trunk becomes smooth as if turned in a lathe.

After some hours in this charming forest I started on the return journey before night, on account of the difficulties of the swamp and the brier patch. On leaving the palmettos and entering the vine-tangled, half-submerged forest I sought long and carefully, but in vain, for the trail, for I had drifted about too incautiously in search of plants. But, recollecting the direction that I had followed in the morning, I took a compass bearing and started to penetrate the swamp in a direct line.

Of course I had a sore weary time, pushing through the tanglement of falling, standing, and half-fallen trees and bushes, to say nothing of knotted vines as remarkable for their efficient army of interlocking and lancing prickers as for their length and the number of their blossoms. But these were not my greatest obstacles, nor yet the pools and lagoons full of dead leaves and alligators. It was the army of cat-briers that I most dreaded. I knew that I would have to find the narrow slit of a lane before dark or spend the night with mosquitoes and alligators, without food or fire. The entire distance was not great, but a traveler in open woods can form no idea of the crooked and strange difficulties of pathless locomotion in these thorny, watery Southern tangles, especially in pitch darkness. I struggled hard and kept my course, leaving the general direction only when drawn aside by a plant of extraordinary promise, that I wanted for a specimen, or when I had to make the half-circuit of a pile of trees, or of a deep lagoon or pond.

In wading I never attempted to keep my clothes dry, because the water was too deep, and the necessary care would consume too much time. Had the water that I was forced to wade been transparent it would have lost much of its difficulty. But as it was, I constantly expected to plant my feet on an alligator, and therefore proceeded with strained caution. The opacity of the water caused uneasiness also on account of my inability to determine its depth. In many places I was compelled to turn back, after wading forty or fifty yards, and to try again a score of times before I succeeded in getting across a single lagoon.

At length, after miles of wading and wallowing, I arrived at the grand cat-brier encampment which guarded the whole forest in solid phalanx, unmeasured miles up and down across my way. Alas! the trail by which I had crossed in the morning was not to be found, and night was near. In vain I scrambled back and forth in search of an opening. There was not even a strip of dry ground on which to rest. Everywhere the long briers arched over to the vines and bushes of the watery swamp, leaving no standing-ground between them. I began to think of building some sort of a scaffold in a tree to rest on through the night, but concluded to make one more desperate effort to find the narrow track.

After calm, concentrated recollection of my course, I made a long exploration toward the left down the brier line, and after scrambling a mile or so, perspiring and bleeding, I discovered the blessed trail and escaped to dry land and the light. Reached the captain at sundown. Dined on milk and johnny-cake and fresh venison. Was congratulated on my singular good fortune and woodcraft, and soon after supper was sleeping the deep sleep of the weary and the safe.

October 22. This morning I was easily prevailed upon by the captain and an ex-judge, who was rusticating here, to join in a deer hunt. Had a delightful ramble in the long grass and flowery barrens. Started one deer but did not draw a single shot. The captain, the judge, and myself stood at different stations where the deer was expected to pass, while a brother of the captain entered the woods to arouse the game from cover. The one deer that he started took a direction different from any which this particular old buck had ever been known to take in times past, and in so doing was cordially cursed as being the "d——dest deer that ever ran unshot." To me it appeared as "d——dest" work to slaughter God's cattle for sport. "They were made for us," say these self-approving preachers; "for our food, our recreation, or other uses not yet discovered." As truthfully we might say on behalf of a bear, when he deals successfully with an unfortunate hunter, "Men and other bipeds were made for bears, and thanks be to God for claws and teeth so long."

Let a Christian hunter go to the Lord's woods and kill his well-kept beasts, or wild Indians, and it is well; but let an enterprising specimen of these proper, predestined victims go to houses and fields and kill the most worthless person of the vertical godlike killers,—oh! that is horribly unorthodox, and on the part of the Indians atrocious murder! Well, I have precious little sympathy for the selfish propriety of civilized man, and if a war of races should occur between the wild beasts and Lord Man, I would be tempted to sympathize with the bears.

CHAPTER VI

CEDAR KEYS

OCTOBER 23. To-day I reached the sea. While I was yet many miles back in the palmy woods, I caught the scent of the salt sea breeze which, although I had so many years lived far from sea breezes, suddenly conjured up Dunbar, its rocky coast, winds and waves; and my whole childhood, that seemed to have utterly vanished in the New World, was now restored amid the Florida woods by that one breath from the sea. Forgotten were the palms and magnolias and the thousand flowers that enclosed me. I could see only dulse and tangle, long-winged gulls, the Bass Rock in the Firth of Forth, and the old castle, schools, churches, and long country rambles in search of birds' nests. I do not wonder that the weary camels coming from the scorching African deserts should be able to scent the Nile.

How imperishable are all the impressions that ever vibrate one's life! We cannot forget anything. Memories may escape the action of will, may sleep a long time, but when stirred by the right influence, though that influence be light as a shadow, they flash into full stature and life with everything in place. For nineteen years my vision was bounded by forests, but to-day, emerging from a multitude of tropical plants, I beheld the Gulf of Mexico stretching away unbounded, except by the sky. What dreams and speculative matter for thought arose as I stood on the strand, gazing out on the burnished, treeless plain!

But now at the seaside I was in difficulty. I had reached a point that I could not ford, and Cedar Keys had an empty harbor. Would I proceed down the peninsula to Tampa and Key West, where I would be sure to find a vessel for Cuba, or would I wait here, like Crusoe, and pray for a ship. Full of these thoughts, I stepped into a little store which had a considerable trade in quinine and alligator and rattlesnake skins, and inquired about shipping, means of travel, etc.

The proprietor informed me that one of several sawmills near the village was running, and that a schooner chartered to carry a load of lumber to Galveston, Texas, was expected at the mills for a load. This mill was situated on a tongue of land a few miles along the coast from Cedar Keys, and I determined to see Mr. Hodgson, the owner, to find out particulars about the expected schooner, the time she would take to load, whether I would be likely to obtain passage on her, etc.

Found Mr. Hodgson at his mill. Stated my case, and was kindly furnished the desired information. I determined to wait the two weeks likely to elapse before she sailed, and go on her to the flowery plains of Texas, from any of whose ports, I fancied, I could easily find passage to the West Indies. I agreed to work for Mr. Hodgson in the mill until I sailed, as I had but little money. He invited me to his spacious house, which occupied a shell hillock and commanded a fine view of the Gulf and many gems of palmy islets, called "keys," that fringe the shore like huge bouquets — not too big, however, for the spacious waters. Mr. Hodgson's family welcomed me with that open, unconstrained cordiality which is characteristic of the better class of Southern people.

At the sawmill a new cover had been put on the main driving pulley, which, made of rough plank, had to be turned off and smoothed. He asked me if I was able to do this job and I told him that I could. Fixing a rest and making a tool out of an old file, I directed the engineer to start the engine and run slow. After turning down the pulley and getting it true, I put a keen edge on a common carpenter's plane, quickly finished the job, and was assigned a bunk in one of the employees' lodging-houses.

The next day I felt a strange dullness and headache while I was botanizing along the coast. Thinking that a bath in the salt water might refresh me, I plunged in and swam a little distance, but this seemed only to make me feel worse. I felt anxious for something sour, and walked back to the village to buy lemons.

Thus and here my long walk was interrupted. I thought that a few days' sail would land me among the famous flower-beds of Texas. But the expected ship came and went while I was helpless with fever. The very day after reaching the sea I began to be weighed down by inexorable leaden numbness, which I resisted and tried to shake off for three days, by bathing in the Gulf, by dragging myself about among the palms, plants, and strange shells of the shore, and by doing a little mill work. I did not fear any serious illness, for I never was sick before, and was unwilling to pay attention to my feelings.

But yet heavier and more remorselessly pressed the growing fever, rapidly gaining on my strength. On the third day after my arrival I could not take any nourishment, but craved acid. Cedar Keys was only a mile or two distant, and I managed to walk there to buy lemons. On returning, about the middle of the afternoon, the fever broke on me like a storm, and before I had staggered halfway to the mill I fell down unconscious on the narrow trail among dwarf palmettos.

When I awoke from the hot fever sleep, the stars were shining, and I was at a loss to know which end of the trail to take, but fortunately, as it afterwards proved, I guessed right. Subsequently, as I fell again and again after walking only a hundred yards or so, I was careful to lie with my head in the direction in which I thought the mill was. I rose, staggered, and fell, I know not how many times, in delirious bewilderment, gasping and throbbing with only moments of consciousness. Thus passed the hours till after midnight, when I reached the mill lodging-house.

The watchman on his rounds found me lying on a heap of sawdust at the foot of the stairs. I asked him to assist me up the steps to bed, but he thought my difficulty was only intoxication and refused to help me. The mill hands, especially on Saturday nights, often returned from the village drunk. This was the cause of the watchman's refusal. Feeling that I must get to bed, I made out to reach it on hands and knees, tumbled in after a desperate struggle, and immediately became oblivious to everything.

I awoke at a strange hour on a strange day to hear Mr. Hodgson ask a watcher beside me whether I had yet spoken, and when he replied that I had not, he said: "Well, you must keep on pouring in quinine. That's all we can do." How long I lay unconscious I never found out, but it must have been many days. Some time or other I was moved on a horse from the mill quarters to Mr. Hodgson's house, where I was nursed about three months with unfailing kindness, and to the skill and care of Mr. and Mrs. Hodgson I doubtless owe my life. Through quinine and calomel — in sorry abundance — with other milder medicines, my malarial fever became typhoid. I had night sweats, and my legs became like posts of the temper and consistency of clay on account of dropsy. So on until January, a weary time.

As soon as I was able to get out of bed, I crept away to the edge of the wood, and sat day after day beneath a moss-draped live-oak, watching birds feeding on the shore when the tide was out. Later, as I gathered some strength, I sailed in a little skiff from one key to another. Nearly all the shrubs and trees here are evergreen, and a few of the smaller plants are in flower all winter. The principal trees on this Cedar Key are the juniper, long-leafed pine, and live-oak. All of the latter, living and dead, are heavily draped with tillandsia, like those of Bonaventure. The leaf is oval, about two inches long, three fourths of an inch wide, glossy and dark green above, pale beneath. The trunk is usually much divided, and is extremely unwedgeable. The specimen on the opposite page[1] is growing in the dooryard of Mr. Hodgson's house. It is a grand old king, whose crown gleamed in the bright sky long ere the Spanish shipbuilders felled a single tree of this noble species.

[1] Of the original journal.

The live-oaks of these keys divide empire with the long-leafed pine and palmetto, but in many places on the mainland there are large tracts exclusively occupied by them. Like the Bonaventure oaks they have the upper side of their main spreading branches thickly planted with ferns, grasses, small saw palmettos, etc. There is also a dwarf oak here, which forms dense thickets. The oaks of this key are not, like those of the Wisconsin openings, growing on grassy slopes, but stand, sunk to the shoulders, in flowering magnolias, heathworts, etc.

During my long sojourn here as a convalescent I used to lie on my back for whole days beneath the ample arms of these great trees, listening to the winds and the birds. There is an extensive shallow on the coast, close by, which the receding tide exposes daily. This is the feeding-ground of thousands of waders of all sizes, plumage, and language, and they make a lively picture and noise when they gather at the great family board to eat their daily bread, so bountifully provided for them.

Their leisure in time of high tide they spend in various ways and places. Some go in large flocks to reedy margins about the islands and wade and stand about quarrelling or making sport, occasionally finding a stray mouthful to eat. Some stand on the mangroves of the solitary shore, now and then plunging into the water after a fish. Some go long journeys inland, up creeks and inlets. A few lonely old herons of solemn look and wing retire to favorite oaks. It was my delight to watch those old white sages of immaculate feather as they stood erect drowsing away the dull hours between tides, curtained by long skeins of tillandsia. White-bearded hermits gazing dreamily from dark caves could not appear more solemn or more becomingly shrouded from the rest of their fellow beings.

One of the characteristic plants of these keys is the Spanish bayonet, a species of yucca, about eight or ten feet in height, and with a trunk three or four inches in diameter when full grown. It belongs to the lily family and develops palmlike from terminal buds. The stout leaves are very rigid, sharp-pointed and bayonet-like. By one of these leaves a man might be as seriously stabbed as by an army bayonet, and woe to the luckless wanderer who dares to urge his way through these armed gardens after dark. Vegetable cats of many species will rob him of his clothes and claw his flesh, while dwarf palmettos will saw his bones, and the bayonets will glide to his joints and marrow without the smallest consideration for Lord Man.

The climate of these precious islets is simply warm summer and warmer summer, corresponding in time with winter and summer in the North. The weather goes smoothly over the points of union betwixt the twin summers. Few of the storms are very loud or variable. The average temperature during the day, in December, was about sixty-five degrees in the shade, but on one day a little damp snow fell.

Cedar Key is two and one half or three miles in diameter and its highest point is forty-four feet above mean tide-water. It is surrounded by scores of other keys, many of them looking like a clump of palms, arranged like a tasteful bouquet, and placed in the sea to be kept fresh. Others have quite a sprinkling of oaks and junipers, beautifully united with vines. Still others consist of shells, with a few grasses and mangroves, circled with a rim of rushes. Those which have sedgy margins furnish a favorite retreat for countless waders and divers, especially for the pelicans that frequently whiten the shore like a ring of foam.

It is delightful to observe the assembling of these feathered people from the woods and reedy isles; herons white as wave-tops, or blue as the sky, winnowing the warm air on wide quiet wing; pelicans coming with baskets to fill, and the multitude of smaller sailors of the air, swift as swallows, gracefully taking their places at Nature's family table for their daily bread. Happy birds!

The mockingbird is graceful in form and a fine singer, plainly dressed, rather familiar in

LIME KEY, FLORIDA
From Mr. Muir's sketch in the original journal

habits, frequently coming like robins to door-sills for crumbs — a noble fellow, beloved by everybody. Wild geese are abundant in winter, associated with brant, some species of which I have never seen in the North. Also great flocks of robins, mourning doves, bluebirds, and the delightful brown thrashers. A large number of the smaller birds are fine singers. Crows, too, are here, some of them cawing with a foreign accent. The common bob-white quail I observed as far south as middle Georgia.

Lime Key, sketched on the opposite page, is a fair specimen of the Florida keys on this part of the coast. A fragment of cactus, *Opuntia*, sketched on another page,[1] is from the above-named key, and is abundant there. The fruit, an inch in length, is gathered, and made into a sauce, of which some people are fond. This species forms thorny, impenetrable thickets. One joint that I measured was fifteen inches long.

The mainland of Florida is less salubrious

[1] Of the original journal.

than the islands, but no portion of this coast, nor of the flat border which sweeps from Maryland to Texas, is quite free from malaria. All the inhabitants of this region, whether black or white, are liable to be prostrated by the ever-present fever and ague, to say nothing of the plagues of cholera and yellow fever that come and go suddenly like storms, prostrating the population and cutting gaps in it like hurricanes in woods.

The world, we are told, was made especially for man — a presumption not supported by all the facts. A numerous class of men are painfully astonished whenever they find anything, living or dead, in all God's universe, which they cannot eat or render in some way what they call useful to themselves. They have precise dogmatic insight of the intentions of the Creator, and it is hardly possible to be guilty of irreverence in speaking of *their* God any more than of heathen idols. He is regarded as a civilized, law-abiding gentleman in favor either of a republican form of government or of a limited monarchy; believes in the literature and language of England; is a warm supporter of the English constitution and Sunday schools and missionary societies; and is as purely a manufactured article as any puppet of a halfpenny theater.

With such views of the Creator it is, of course, not surprising that erroneous views should be entertained of the creation. To such properly trimmed people, the sheep, for example, is an easy problem — food and clothing "for us," eating grass and daisies white by divine appointment for this predestined purpose, on perceiving the demand for wool that would be occasioned by the eating of the apple in the Garden of Eden.

In the same pleasant plan, whales are storehouses of oil for us, to help out the stars in lighting our dark ways until the discovery of the Pennsylvania oil wells. Among plants, hemp, to say nothing of the cereals, is a case of evident destination for ships' rigging, wrapping packages, and hanging the wicked. Cotton is another plain case of clothing. Iron was made for hammers and ploughs, and lead for bullets; all intended for us. And so of other small handfuls of insignificant things.

But if we should ask these profound expositors of God's intentions, How about those man-eating animals — lions, tigers, alligators — which smack their lips over raw man? Or about those myriads of noxious insects that destroy labor and drink his blood? Doubtless man was intended for food and drink for all these? Oh, no! Not at all! These are unresolvable difficulties connected with Eden's apple and the Devil. Why does water drown its lord? Why do so many minerals poison him? Why are so many plants and fishes deadly enemies? Why is the lord of creation subjected to the same laws of life as his subjects? Oh, all these things are satanic, or in some way connected with the first garden.

Now, it never seems to occur to these far-seeing teachers that Nature's object in making animals and plants might possibly be first of all the happiness of each one of them, not the creation of all for the happiness of one. Why should man value himself as more than a small part of the one great unit of creation? And what creature of all that the Lord has taken the pains to make is not essential to the completeness of that unit — the cosmos? The universe would be incomplete without man; but it would also be incomplete without the smallest transmicroscopic creature that dwells beyond our conceitful eyes and knowledge.

From the dust of the earth, from the common elementary fund, the Creator has made *Homo sapiens*. From the same material he has made every other creature, however noxious and insignificant to us. They are earth-born companions and our fellow mortals. The fearfully good, the orthodox, of this laborious patchwork of modern civilization cry "Heresy" on every one whose sympathies reach a single hair's breadth beyond the boundary epidermis of our own species. Not content with taking all of earth, they also claim the celestial country as the only ones who possess the kind of souls for which that imponderable empire was planned.

This star, our own good earth, made many a successful journey around the heavens ere man was made, and whole kingdoms of creatures enjoyed existence and returned to dust ere man appeared to claim them. After human beings have also played their part in Creation's plan, they too may disappear without any general burning or extraordinary commotion whatever.

Plants are credited with but dim and uncertain sensation, and minerals with positively none at all. But why may not even a mineral arrangement of matter be endowed with sensation of a kind that we in our blind exclusive perfection can have no manner of communication with?

But I have wandered from my object. I stated a page or two back that man claimed the earth was made for him, and I was going to say that venomous beasts, thorny plants, and deadly diseases of certain parts of the earth prove that the whole world was not made for him. When an animal from a tropical climate is taken to high latitudes, it may perish of cold, and we say that such an animal was never intended for so severe a climate. But when man betakes himself to sickly parts of the tropics and perishes, he cannot see that he was never intended for such deadly climates. No, he will rather accuse the first mother of the cause of the difficulty, though she may never have seen a fever district; or will consider it a providential chastisement for some self-invented form of sin.

Furthermore, all uneatable and uncivilizable animals, and all plants which carry prickles, are deplorable evils which, according to closet researches of clergy, require the cleansing chemistry of universal planetary combustion. But more than aught else mankind requires burning, as being in great part wicked, and if that transmundane furnace can be so applied and regulated as to smelt and purify us into conformity with the rest of the terrestrial creation, then the tophetization of the erratic genus Homo were a consummation devoutly to be prayed for. But, glad to leave these ecclesiastical fires and blunders, I joyfully return to the immortal truth and immortal beauty of Nature.

CHAPTER VII

A SOJOURN IN CUBA

ONE day in January I climbed to the housetop to get a view of another of the fine sunsets of this land of flowers. The landscape was a strip of clear Gulf water, a strip of sylvan coast, a tranquil company of shell and coral keys, and a gloriously colored sky without a threatening cloud. All the winds were hushed and the calm of the heavens was as profound as that of the palmy islands and their encircling waters. As I gazed from one to another of the palm-crowned keys, enclosed by the sunset-colored dome, my eyes chanced to rest upon the fluttering sails of a Yankee schooner that was threading the tortuous channel in the coral reef leading to the harbor of Cedar Keys. "There," thought I, "perhaps I may sail in that pretty white moth." She proved to be the schooner Island Belle.

One day soon after her arrival I went over the key to the harbor, for I was now strong enough to walk. Some of her crew were ashore after water. I waited until their casks were filled, and went with them to the vessel in their boat. Ascertained that she was ready to sail with her cargo of lumber for Cuba. I engaged passage on her for twenty-five dollars, and asked her sharp-visaged captain when he would sail. "Just as soon," said he, "as we get a north wind. We have had northers enough when we did not want them, and now we have this dying breath from the south."

Hurrying back to the house, I gathered my plants, took leave of my kind friends, and went aboard, and soon, as if to calm the captain's complaints, Boreas came foaming loud and strong. The little craft was quickly trimmed and snugged, her inviting sails spread open, and away she dashed to her ocean home like an exulting war-horse to the battle. Islet after islet speedily grew dim and sank beneath the horizon. Deeper became the blue of the water, and in a few hours all of Florida vanished.

This excursion on the sea, the first one after twenty years in the woods, was of course exceedingly interesting, and I was full of hope, glad to be once more on my journey to the South. Boreas increased in power and the Island Belle appeared to glory in her speed and managed her full-spread wings as gracefully as a sea-bird. In less than a day our norther increased in strength to the storm point. Deeper and wider became the valleys, and yet higher the hills of the round plain of water. The flying jib and gaff topsails were lowered and mainsails close-reefed, and our deck was white with broken wave-tops.

"You had better go below," said the captain. "The Gulf Stream, opposed by this wind, is raising a heavy sea and you will be sick. No landsman can stand this long." I replied that I hoped the storm would be as violent as his ship could bear, that I enjoyed the scenery of such a sea so much that it was impossible to be sick, that I had long waited in the woods for just such a storm, and that, now that the precious thing had come, I would remain on deck and enjoy it. "Well," said he, "if you can stand this, you are the first landsman I ever saw that could."

I remained on deck, holding on by a rope to keep from being washed overboard, and watched the behavior of the Belle as she dared nobly on; but my attention was mostly directed among the glorious fields of foam-topped waves. The wind had a mysterious voice and carried nothing now of the songs of birds or of the rustling of palms and fragrant vines. Its burden was gathered from a stormy expanse of crested waves and briny tangles. I could see no striving in those magnificent wave-motions, no raging; all the storm was apparently inspired with nature's beauty and harmony. Every wave was obedient and harmonious as the smoothest ripple of a forest lake, and after dark all the water was phosphorescent like silver fire, a glorious sight.

Our luminous storm was all too short for me. Cuba's rock-waves loomed above the white waters early in the morning. The sailors, accustomed to detect the faintest land line, pointed out well-known guiding harbor-marks back of the Morro Castle long before I could see them through the flying spray. We sailed landward for several hours, the misty shore becoming gradually more earthlike. A flock of white-plumaged ships was departing from the Havana harbor, or, like us, seeking to enter it. No sooner had our little schooner flapped her sails in the lee of the Castle than she was boarded by a swarm of daintily dressed officials who were good-naturedly and good-gesturedly making all sorts of inquiries, while our busy captain, paying little attention to them, was giving orders to his crew.

The neck of the harbor is narrow and it is seldom possible to sail in to appointed anchorage without the aid of a steam tug. Our captain wished to save his money, but after much profitless tacking was compelled to take the proffered aid of steam, when we soon reached our quiet mid-harbor quarters and dropped anchor among ships of every size from every sea.

I was still four or five hundred yards from land and could determine no plant in sight excepting the long arched leaf banners of the banana and the palm, which made a brave show on the Morro Hill. When we were approaching the land, I observed that in some places it was distinctly yellow, and I wondered while we were yet some miles distant whether the color belonged to the ground or to sheets of flowers. From our harbor home I could now see that the color was plant-gold. On one side of the harbor was a city of these yellow plants; on the other, a city of yellow stucco houses, narrowly and confusedly congregated.

"Do you want to go ashore?" said the captain to me. "Yes," I replied, "but I wish to go to the plant side of the harbor." "Oh, well," he said, "come with me now. There are some fine squares and gardens in the city, full of all

MORRO CASTLE AND ENTRANCE TO HAVANA HARBOR

sorts of trees and flowers. Enjoy these to-day, and some other day we will all go over the Morro Hill with you and gather shells. All kinds of shells are over there; but these yellow slopes that you see are covered only with weeds."

We jumped into the boat and a couple of sailors pulled us to the thronged, noisy wharf. It was Sunday afternoon,[1] the noisiest day of a Havana week. Cathedral bells and prayers in the forenoon, theaters and bull-fight bells and bellowings in the afternoon! Lowly whispered prayers to the saints and the Virgin, followed by shouts of praise or reproach to bulls and matadors! I made free with fine oranges and bananas and many other fruits. Pineapple I had never seen before. Wandered about the narrow streets, stunned with the babel of strange sounds and sights; went gazing, also, among the gorgeously flowered garden squares, and then waited among some boxed merchandise until our captain, detained by busi-

[1] Doubtless January 12, 1868.

ness, arrived. Was glad to escape to our little schooner Belle again, weary and heavy laden with excitement and tempting fruits.

As night came on, a thousand lights starred the great town. I was now in one of my happy dreamlands, the fairest of West India islands. But how, I wondered, shall I be able to escape from this great city confusion? How shall I reach nature in this delectable land? Consulting my map, I longed to climb the central mountain range of the island and trace it through all its forests and valleys and over its summit peaks, a distance of seven or eight hundred miles. But alas! though out of Florida swamps, fever was yet weighing me down, and a mile of city walking was quite exhausting. The weather too was oppressively warm and sultry.

January 16. During the few days since our arrival the sun usually has risen unclouded, pouring down pure gold, rich and dense, for one or two hours. Then islandlike masses of white-edged cumuli suddenly appeared, grew to storm size, and in a few minutes discharged rain in tepid plashing bucketfuls, accompanied with high wind. This was followed by a short space of calm, half-cloudy sky, delightfully fragrant with flowers, and again the air would become hot, thick, and sultry.

This weather, as may readily be perceived, was severe to one so weak and feverish, and after a dozen trials of strength over the Morro Hill and along the coast northward for shells and flowers, I was sadly compelled to see that no enthusiasm could enable me to walk to the interior. So I was obliged to limit my researches to within ten or twelve miles of Havana. Captain Parsons offered his ship as my headquarters, and my weakness prevented me from spending a single night ashore.

The daily programme for nearly all the month that I spent here was about as follows: After breakfast a sailor rowed me ashore on the north side of the harbor. A few minutes' walk took me past the Morro Castle and out of sight of the town on a broad cactus common, about as solitary and untrodden as the tangles of

Florida. Here I zigzagged and gathered prizes among unnumbered plants and shells along the shore, stopping to press the plant specimens and to rest in the shade of vine-heaps and bushes until sundown. The happy hours stole away until I had to return to the schooner. Either I was seen by the sailors who usually came for me, or I hired a boat to take me back. Arrived, I reached up my press and a big handful of flowers, and with a little help climbed up the side of my floating home.

Refreshed with supper and rest, I recounted my adventures in the vine tangles, cactus thickets, sunflower swamps, and along the shore among the breakers. My flower specimens, also, and pocketfuls of shells and corals had to be reviewed. Next followed a cool, dreamy hour on deck amid the lights of the town and the various vessels coming and departing.

Many strange sounds were heard: the vociferous, unsmotherable bells, the heavy thundering of cannon from the Castle, and the shouts of the sentinels in measured time. Combined they made the most incessant sharp-angled mass of noise that I ever was doomed to hear. Nine or ten o'clock found me in a small bunk with the harbor wavelets tinkling outside close to my ear. The hours of sleep were filled with dreams of heavy heat, of fruitless efforts for the disentanglement of vines, or of running from curling breakers back to the Morro, etc. Thus my days and nights went on.

Occasionally I was persuaded by the captain to go ashore in the evening on his side of the harbor, accompanied perhaps by two or three other captains. After landing and telling the sailors when to call for us, we hired a carriage and drove to the upper end of the city, to a fine public square adorned with shady walks and magnificent plants. A brass band in imposing uniform played the characteristic lance-noted martial airs of the Spanish. Evening is the fashionable hour for aristocratic drives about the streets and squares, the only time that is delightfully cool. I never saw elsewhere people so neatly and becomingly dressed. The proud best-family Cubans may fairly be called beautiful, are under- rather than over-sized, with features exquisitely moulded, and set off with silks and broadcloth in excellent taste. Strange that their amusements should be so coarse. Bull-fighting, brain-splitting bell-ringing, and the most piercing artificial music appeal to their taste.

The rank and wealth of Havana nobility, when out driving, seems to be indicated by the distance of their horses from the body of the carriage. The higher the rank, the longer the shafts of the carriage, and the clumsier and more ponderous are the wheels, which are not unlike those of a cannon-cart. A few of these carriages have shafts twenty-five feet in length, and the brilliant-liveried negro driver on the lead horse, twenty or thirty feet in advance of the horse in the shafts, is beyond calling distance of his master.

Havana abounds in public squares, which in all my random strolls throughout the big town

I found to be well watered, well cared for, well planted, and full of exceedingly showy and interesting plants, rare even amid the exhaustless luxuriance of Cuba. These squares also contained fine marble statuary and were furnished with seats in the shadiest places. Many of the walks were paved instead of graveled.

The streets of Havana are crooked, labyrinthic, and exceedingly narrow. The sidewalks are only about a foot wide. A traveler experiences delightful relief when, heated and wearied by raids through the breadth of the dingy yellow town, dodging a way through crowds of men and mules and lumbering carts and carriages, he at length finds shelter in the spacious, dustless, cool, flowery squares; still more when, emerging from all the din and darkness of these lanelike streets, he suddenly finds himself out in the middle of the harbor, inhaling full-drawn breaths of the sea breezes.

The interior of the better houses which came under my observation struck me with the profusion of dumpy, ill-proportioned pillars at the entrances and in the halls, and with the spacious open-fielded appearance of their enclosed square house-gardens or courts. Cubans in general appear to me superfinely polished, polite, and agreeable in society, but in their treatment of animals they are cruel. I saw more downright brutal cruelty to mules and horses during the few weeks I stayed there than in my whole life elsewhere. Live chickens and hogs are tied in bunches by the legs and carried to market thus, slung on a mule. In their general treatment of all sorts of animals they seem to have no thought for them beyond cold-blooded, selfish interest.

In tropical regions it is easy to build towns, but it is difficult to subdue their armed and united plant inhabitants, and to clear fields and make them blossom with breadstuff. The plant people of temperate regions, feeble, unarmed, unallied, disappear under the trampling feet of flocks, herds, and man, leaving their homes to enslavable plants which follow the will of man and furnish him with food. But the armed and united plants of the tropics hold their rightful kingdom plantfully, nor, since the first appearance of Lord Man, have they ever suffered defeat.

A large number of Cuba's wild plants circle closely about Havana. In five minutes' walk from the wharf I could reach the undisturbed settlements of Nature. The field of the greater portion of my rambling researches was a strip of rocky common, silent and unfrequented by anybody save an occasional beggar at Nature's door asking a few roots and seeds. This natural strip extended ten miles along the coast northward, with but few large-sized trees and bushes, but rich in magnificent vines, cacti-composites, leguminous plants, grasses, etc. The wild flowers of this seaside field are a happy band, closely joined in splendid array. The trees shine with blossoms and with light reflected from the leaves. The individuality of the vines is lost in trackless, interlacing, twisting, overheaping union.

Our American "South" is rich in flowery

vines. In some districts almost every tree is crowned with them, aiding each other in grace and beauty. Indiana, Kentucky, and Tennessee have the grapevine in predominant numbers and development. Farther south dwell the greenbriers and countless leguminous vines. A vine common among the Florida islets, perhaps belonging to the dogbane family, overruns live-oaks and palmettos, with frequently more than a hundred stems twisted into one cable. Yet in no section of the South are there such complicated and such gorgeously flowered vine-tangles as flourish in armed safety in the hot and humid wild gardens of Cuba.

The longest and the shortest vine that I found in Cuba were both leguminous. I have said that the harbor side of the Morro Hill is clothed with tall yellow-flowered composites through which it is difficult to pass. But there are smooth, velvety, lawnlike patches in these *Compositæ* forests. Coming suddenly upon one of these open places, I stopped to admire its greenness and smoothness, when I observed a sprinkling of large papilionaceous blossoms among the short green grass. The long composites that bordered this little lawn were entwined and almost smothered with vines which bore similar corollas in tropic abundance.

I at once decided that these sprinkled flowers had been blown off the encompassing tangles and had been kept fresh by dew and by spray from the sea. But, on stooping to pick one of them up, I was surprised to find that it was attached to Mother Earth by a short, prostrate, slender hair of a vine stem, bearing, besides the one large blossom, a pair or two of linear leaves. The flower weighed more than stem, root, and leaves combined. Thus, in a land of creeping and twining giants, we find also this charming, diminutive simplicity — the vine reduced to its lowest terms.

The longest vine, prostrate and untwined like its little neighbor, covers patches of several hundred square yards with its countless branches and close growth of upright, trifoliate, smooth green leaves. The flowers are as plain and unshowy in size and color as those of the sweet peas of gardens. The seeds are large and satiny. The whole plant is noble in its motions and features, covering the ground with a depth of unconfused leafage which I have never seen equaled by any other plant. The extent of leaf-surface is greater, I think, than that of a large Kentucky oak. It grows, as far as my observation has reached, only upon shores, in a soil composed of broken shells and corals, and extends exactly to the water-line of the highest-reaching waves. The same plant is abundant in Florida.

The cacti form an important part of the plant population of my ramble ground. They are various as the vines, consisting now of a diminutive joint or two hid in the weeds, now rising into bushy trees, wide-topped, with trunks a foot in diameter, and with glossy, dark-green joints that reflect light like the silex-varnished palms. They are planted for fences, together with the Spanish bayonet and agave.

In one of my first walks I was laboriously

scrambling among some low rocks gathering ferns and vines, when I was startled by finding my face close to a great snake, whose body was disposed carelessly like a castaway rope among the weeds and stones. After escaping and coming to my senses, I discovered that the snake was a member of the vegetable kingdom, capable of no dangerous amount of locomotion, but possessed of many a fang, and prostrate as though under the curse of Eden, "Upon thy belly shalt thou go and dust shalt thou eat."

One day, after luxuriating in the riches of my Morro pasture, and pressing many new specimens, I went down to the bank of brilliant wave-washed shells to rest awhile in their beauty, and to watch the breakers that a powerful norther was heaving in splendid rank along the coral boundary. I gathered pocketfuls of shells, mostly small but fine in color and form, and bits of rosy coral. Then I amused myself by noting the varying colors of the waves and the different forms of their curved and blossoming crests. While thus alone and free it was interesting to learn the richly varied songs, or what we mortals call the roar, of expiring breakers. I compared their variation with the different distances to which the broken wave-water reached landward in its farthest-flung foam-wreaths, and endeavored to form some idea of the one great song sounding forever all around the white-blooming shores of the world.

Rising from my shell seat, I watched a wave leaping from the deep and coming far up the beveled strand to bloom and die in a mass of white. Then I followed the spent waters in their return to the blue deep, wading in their spangled, decaying fragments until chased back up the bank by the coming of another wave. While thus playing half studiously, I discovered in the rough, beaten deathbed of the wave a little plant with closed flowers. It was crouching in a hollow of the brown wave-washed rock, and one by one the chanting, dying waves rolled over it. The tips of its delicate pink petals peered above the clasping green calyx. "Surely," said I, as I stooped over it for a moment, before the oncoming of another wave, "surely you cannot be living here! You must have been blown from some warm bank, and rolled into this little hollow crack like a dead shell." But, running back after every retiring wave, I found that its roots were wedged into a shallow wrinkle of the coral rock, and that this wave-beaten chink was indeed its dwelling-place.

I had oftentimes admired the adaptation displayed in the structure of the stately dulse and other seaweeds, but never thought to find a highbred flowering plant dwelling amid waves in the stormy, roaring domain of the sea. This little plant has smooth globular leaves, fleshy and translucent like beads, but green like those of other land plants. The flower is about five eighths of an inch in diameter, rose-purple, opening in calm weather, when deserted by the waves. In general appearance it is like a small portulaca. The strand, as far as I walked it, was luxuriantly fringed with woody *Compositæ*, two or three feet in height, their tops purple and golden with a profusion of flowers. Among these I discovered a small bush whose yellow flowers were ideal; all the parts were present regularly alternate and in fives, and all separate, a plain harmony.

When a page is written over but once it may be easily read; but if it be written over and over with characters of every size and style, it soon becomes unreadable, although not a single confused meaningless mark or thought may occur among all the written characters to mar its perfection. Our limited powers are similarly perplexed and overtaxed in reading the inexhaustible pages of nature, for they are written over and over uncountable times, written in characters of every size and color, sentences composed of sentences, every part of a character a sentence. There is not a fragment in all nature, for every relative fragment of one thing is a full harmonious unit in itself. All together form the one grand palimpsest of the world.

One of the most common plants of my pasture was the agave. It is sometimes used for fencing. One day, in looking back from the top of the Morro Hill, as I was returning to the Island Belle, I chanced to observe two poplar-like trees about twenty-five feet in height. They were growing in a dense patch of cactus and vine-knotted sunflowers. I was anxious to see anything so homelike as a poplar, and so made haste towards the two strange trees, making a way through the cactus and sunflower jungle that protected them. I was surprised to find that what I took to be poplars were agaves in flower, the first I had seen. They were almost out of flower, and fast becoming wilted at the approach of death. Bulbs were scattered about, and a good many still remained on the branches, which gave it a fruited appearance.

The stem of the agave seems enormous in size when one considers that it is the growth of a few weeks. This plant is said to make a mighty effort to flower and mature its seeds and then to die of exhaustion. Now there is not, so far as I have seen, a mighty effort or the need of one, in wild Nature. She accomplishes her ends without unquiet effort, and perhaps there is nothing more mighty in the development of the flower-stem of the agave than in the development of a grass panicle.

Havana has a fine botanical garden. I spent pleasant hours in its magnificent flowery arbors and around its shady fountains. There is a palm avenue which is considered wonderfully stately and beautiful, fifty palms in two straight lines, each rigidly perpendicular. The smooth round shafts, slightly thicker in the middle, appear to be productions of the lathe, rather than vegetable stems. The fifty arched crowns, inimitably balanced, blaze in the sunshine like heaps of stars that have fallen from the skies. The stems were about sixty or seventy feet in height, the crowns about fifteen feet in diameter.

Along a stream-bank were tall, waving bamboos, leafy as willows, and infinitely graceful in wind gestures. There was one species of palm, with immense bipinnate leaves and leaflets fringed, jagged, and one-sided, like those of *Adiantum*. Hundreds of the most gorgeous-flowered plants, some of them large trees, belonging to the *Leguminosæ*. Compared with what I have before seen in artificial flower-gardens, this is past comparison the grandest. It is a perfect metropolis of the brightest and most exuberant of garden plants, watered by handsome fountains, while graveled and finely bordered walks slant and curve in all directions, and in all kinds of fanciful playground styles, more like the fairy gardens of the Arabian Nights than any ordinary man-made pleasure-ground.

In Havana I saw the strongest and the ugliest negroes that I have met in my whole walk. The stevedores of the Havana wharf are muscled in true giant style, enabling them to tumble and toss ponderous casks and boxes of sugar weighing hundreds of pounds as if they were empty. I heard our own brawny sailors, after watching them at work a few minutes, express unbounded admiration of their strength, and wish that their hard outbulging muscles were for sale. The countenances of some of the negro orange-selling dames express a devout good-natured ugliness that I never could have conceived any arrangement of flesh and blood to be capable of. Besides oranges they sold pineapples, bananas, and lottery tickets.

CHAPTER VIII

BY A CROOKED ROUTE TO CALIFORNIA

AFTER passing a month in this magnificent island, and finding that my health was not improving, I made up my mind to push on to South America while my stock of strength, such as it was, lasted. But fortunately I could not find passage for any South American port. I had long wished to visit the Orinoco basin and in particular the basin of the Amazon. My plan was to get ashore anywhere on the north end of the continent, push on southward through the wilderness around the headwaters of the Orinoco, until I reached a tributary of the Amazon, and float down on a raft or skiff the whole length of the great river to its mouth. It seems strange that such a trip should ever have entered the dreams of any person, however enthusiastic and full of youthful daring, particularly under the disadvantages of poor health, of funds less than a hundred dollars, and of the insalubrity of the Amazon Valley.

Fortunately, as I said, after visiting all the shipping agencies, I could not find a vessel of any sort bound for South America, and so made up a plan to go North, to the longed-for cold weather of New York, and thence to the forests and mountains of California. There, I thought, I shall find health and new plants and mountains, and after a year spent in that interesting country I can carry out my Amazon plans.

It seemed hard to leave Cuba thus unseen and unwalked, but illness forbade my stay and I had to comfort myself with the hope of returning to its waiting treasures in full health. In the mean time I prepared for immediate departure. When I was resting in one of the Havana gardens, I noticed in a New York paper an advertisement of cheap fares to California. I consulted Captain Parsons concerning a passage to New York, where I could find a ship for California. At this time none of the California ships touched at Cuba.

"Well," said he, pointing toward the middle of the harbor, "there is a trim little schooner loaded with oranges for New York, and these little fruiters are fast sailers. You had better see her captain about a passage, for she must be about ready to sail." So I jumped into the dinghy and a sailor rowed me over to the fruiter. Going aboard, I inquired for the captain, who soon appeared on deck and readily agreed to carry me to New York for twenty-five dollars. Inquiring when he would sail, "To-morrow morning at daylight," he replied, "if this norther slacks a little; but my papers are made out, and you will have to see the American consul to get permission to leave on my ship."

I immediately went to the city, but was unable to find the consul, whereupon I determined to sail for New York without any formal leave. Early next morning, after leaving the Island Belle and bidding Captain Parsons good-bye, I was rowed to the fruiter and got aboard. Notwithstanding the north wind was still as boisterous as ever, our Dutch captain was resolved to face it, confident in the strength of his all-oak little schooner.

Vessels leaving the harbor are stopped at the Morro Castle to have their clearance papers examined; in particular, to see that no runaway slaves were being carried away. The officials came alongside our little ship, but did not come aboard. They were satisfied by a glance at the consul's clearance paper, and with the declaration of the captain, when asked whether he had any negroes, that he had "not a d——d one." "All right, then," shouted the officials, "farewell! A pleasant voyage to you!" As my name was not on the ship's papers, I stayed below, out of sight, until I felt the heaving of the waves and knew that we were fairly out on the open sea. The Castle towers, the hills, the palms, and the wave-white strand, all faded in the distance, and our mimic sea-bird was at home in the open stormy gulf, curtsying to every wave and facing bravely to the wind.

Two thousand years ago our Saviour told Nicodemus that he did not know where the winds came from, nor where they were going. And now in this Golden Age, though we Gentiles know the birthplace of many a wind and also "whither it is going," yet we know about as little of winds in general as those Palestinian Jews, and our ignorance, despite the powers of science, can never be much less profound than it is at present.

The substance of the winds is too thin for human eyes, their written language is too difficult for human minds, and their spoken language mostly too faint for the ears. A mechanism is said to have been invented whereby the human organs of speech are made to write their own utterances. But without any extra mechanical contrivance, every speaker also writes as he speaks. All things in the creation of God register their own acts. The poet was mistaken when he said, "From the wing no scar the sky sustains." His eyes were simply too dim to see the scar. In sailing past Cuba I could see a fringe of foam along the coast, but could hear no sound of waves, simply because my ears could not hear wave-dashing at that distance. Yet every bit of spray was sounding in my ears.

The subject brings to mind a few recollections of the winds I heard in my late journey. In my walk from Indiana to the Gulf, earth and sky, plants and people, and all things changeable were constantly changing. Even in Kentucky nature and art have many a characteristic shibboleth. The people differ in language and in customs. Their architecture is generically different from that of their immediate neighbors on the north, not only in planters' mansions, but in barns and granaries and the cabins of the poor. But thousands of familiar flower faces looked from every hill and valley. I noted no difference in the sky, and the winds spoke the same things. I did not feel myself in a strange land.

In Tennessee my eyes rested upon the first mountain scenery I ever beheld. I was rising higher than ever before; strange trees were beginning to appear; alpine flowers and shrubs were meeting me at every step. But these Cumberland Mountains were timbered with oak, and were not unlike Wisconsin hills piled upon each other, and the strange plants were like those that were not strange. The sky was changed only a little, and the winds not by a single detectible note. Therefore, neither was Tennessee a strange land.

But soon came changes thick and fast. After passing the mountainous corner of North Carolina and a little way into Georgia, I beheld from one of the last ridge-summits of the Alleghanies that vast, smooth, sandy slope that reaches from the mountains to the sea. It is wooded with dark, branchy pines which were all strangers to me. Here the grasses, which are an earth-covering at the North, grow wide apart in tall clumps and tufts like saplings. My known flower companions were leaving me now, not one by one as in Kentucky and Tennessee, but in whole tribes and genera, and companies of shining strangers came trooping upon me in countless ranks. The sky, too, was changed, and I could detect strange sounds in the winds. Now I began to feel myself "a stranger in a strange land."

But in Florida came the greatest change of all, for here grows the palmetto, and here blow the winds so strangely toned by them. These palms and these winds severed the last strands of the cord that united me with home. Now I was a stranger, indeed. I was delighted, astonished, confounded, and gazed in wonderment blank and overwhelming as if I had fallen upon another star. But in all of this long, complex series of changes, one of the greatest, and the last of all, was the change I found in the tone and language of the winds. They no longer came with the old home music gathered from open prairies and waving fields of oak, but they passed over many a strange string. The leaves of magnolia, smooth like polished steel, the immense inverted forests of tillandsia banks, and the princely crowns of palms — upon these the winds made strange music, and at the coming-on of night had overwhelming power to present the distance from friends and home, and the completeness of my isolation from all things familiar.

Elsewhere I have already noted that when I was a day's journey from the Gulf, a wind blew upon me from the sea — the first sea breeze that had touched me in twenty years. I was plodding along with my satchel and plants, leaning wearily forward, a little sore from approaching fever, when suddenly I felt the salt air, and before I had time to think, a whole flood of long-dormant associations rolled in upon me. The Firth of Forth, the Bass Rock, Dunbar Castle, and the winds and rocks and hills came upon the wings of that wind, and stood in as clear and sudden light as a landscape flashed upon the view by a blaze of lightning in a dark night.

I like to cling to a small chip of a ship like ours when the sea is rough, and long, comet-tailed streamers are blowing from the curled top of every wave. A big vessel responds awkwardly with mixed gestures to several waves at once, lumbering along like a loose floating island. But our little schooner, buoyant as a gull, glides up one side and down the other of each wave hill in delightful rhythm. As we advanced the scenery increased in grandeur and beauty. The waves heaved higher and grew wider, with corresponding motion. It was delightful to ride over this unsullied country of ever-changing water, and when looking upward from the shallow vales, or abroad over the round expanse from the tops of the wave hills, I almost forgot at times that the glassy, treeless country was forbidden to walkers. How delightful it would be to ramble over it on foot, enjoying the transparent crystal ground, and the music of its rising and falling hillocks, unmarred by the ropes and spars of a ship; to study the plants of these waving plains and their stream-currents; to sleep in wild weather in a bed of phosphorescent wave-foam, or briny scented seaweeds; to see the fishes by night in pathways of phosphorescent light; to walk the glassy plain in calm, with birds and flocks of glittering flying fishes here and there, or by night with every star pictured in its bosom!

But even of the land only a small portion is free to man, and if he, among other journeys on forbidden paths, ventures among the ice lands and hot lands, or up in the air in balloon bubbles, or on the ocean in ships, or down into it a little way in smothering diving-bells — in all such small adventures man is admonished and often punished in ways which clearly show him that he is in places for which, to use an approved phrase, he was never designed. However, in view of the rapid advancement of our time, no one can tell how far our star may finally be subdued to man's will. At all events I enjoyed this drifting locomotion to some extent.

The tar-scented community of a ship is a study in itself — a despotism on the small territory of a few drifting planks pinned together. But as our crew consisted only of four sailors, a mate, and the captain, there were no signs of despotism. We all dined at one table, enjoying our fine store of salt mackerel and plum duff, with endless abundance of oranges. Not only was the hold of our little ship filled with loose, unboxed oranges, but the deck also was filled up level with the rails, and we had to walk over the top of the golden fruit on boards.

Flocks of flying fishes often flew across the ship, one or two occasionally falling among the oranges. These the sailors were glad to capture to sell in New York as curiosities, or to give away to friends. But the captain had a large Newfoundland dog who got the largest share of these unfortunate fishes. He used to jump from a dozing sleep as soon as he heard the fluttering of their wings, then pounce and feast leisurely on them before the sailors could reach the spot where they fell.

In passing through the Straits of Florida the winds died away and the sea was smoothed to unruffled calm. The water here is very transparent and of delightfully pure pale-blue color, as different from ordinary dull-colored water as town smoke from mountain air. I could see the bottom as distinctly as one sees the ground when riding over it. It seemed strange that our ship should be upborne in such an ethereal liquid as this, and that we did not run aground where the bottom seemed so near.

One morning, while among the Bahama dots of islands, we had calm sky and calm sea. The sun had risen in cloudless glory, when I observed a large flock of flying fish, a short distance from us, closely pursued by a dolphin. These fish-swallows rose in pretty good order, skimmed swiftly ahead for fifty or a hundred yards in a low arc, then dipped below the surface. Dripping and sparkling, they rose again in a few seconds and glanced back into the lucid brine with wonderful speed, but without apparent terror.

At length the dolphin, gaining on the flock, dashed into the midst of them, and now all order was at an end. They rose in scattering disorder, in all directions, like a flock of birds charged by a hawk. The pursuing dolphin also leaped into the air, showing his splendid colors and wonderful speed. After the first scattering flight all steady pursuit was useless, and the dolphin had but to pounce about in the broken mob of its weary prey until satisfied with his meal.

We are apt to look out on the great ocean and regard it as but a half-blank part of our globe — a sort of desert, "a waste of water." But, land animals though we be, land is about as unknown to us as the sea, for the turbid glances we gain of the ocean in general through commercial eyes are comparatively worthless. Now that science is making comprehensive surveys of the life of the sea, and the forms of its basins, and similar surveys are being made into the land deserts, hot and cold, we may at length discover that the sea is as full of life as the land. None can tell how far man's knowledge may yet reach.

After passing the Straits and sailing up the coast, when about opposite the south end of the Carolina coast, we had stiff head winds all the way to New York and our able little vessel was drenched all day long. Of course our load of oranges suffered, and since they were boarded over level with the rail, we had difficulty in walking and had many chances of being washed overboard. The flying fishes off Cape Hatteras appeared to take pleasure in shooting across from wave-top to wave-top. They avoided the ship during the day, but frequently fell among the oranges at night. The sailors caught many, but our big Newfoundland dog jumped for them faster than the sailors, and so almost monopolized the game.

When dark night fell on the stormy sea, the breaking waves of phosphorescent light were a glorious sight. On such nights I stood on the bowsprit holding on by a rope for hours in order to enjoy this phenomenon. How wonderful this light is! Developed in the sea by myriads of organized beings, it gloriously illuminates the pathways of the fishes, and every breaking wave, and in some places glows over large areas like sheet lightning. We sailed through large fields of seaweed, of which I procured specimens. I thoroughly enjoyed life in this novel little tar-and-oakum home, and, as the end of our voyage drew nigh, I was sorry at the thought of leaving it.

We were now, on the twelfth day, approaching New York, the big ship metropolis. We were in sight of the coast all day. The leafless trees and the snow appeared wonderfully strange. It was now about the end of February and snow covered the ground nearly to the water's edge. Arriving, as we did, in this rough winter weather from the intense heat and general tropical luxuriance of Cuba, the leafless, snow-white woods of New York struck us with all the novelty and impressiveness of a new world. A frosty blast was sweeping seaward from Sandy Hook. The sailors explored their wardrobes for their long-cast-off woolens, and pulled the ropes and managed the sails while muffled in clothing to the rotundity of Eskimos. For myself, long burdened with fever, the frosty wind, as it sifted through my loosened bones, was more delicious and grateful than ever was a spring-scented breeze.

We now had plenty of company; fleets of vessels were on the wing from all countries. Our taut little racer outwinded without exception all who, like her, were going to the port. Toward evening we were grinding and wedging our way through the ice-field of the river delta, which we passed with difficulty. Arrived in port at nine o'clock. The ship was deposited, like a cart at market, in a proper slip, and next morning we and our load of oranges, one third rotten, were landed. Thus all the purposes of our voyage were accomplished.

On our arrival the captain, knowing something of the lightness of my purse, told me that I could continue to occupy my bed on the ship until I sailed for California, getting my meals at a near-by restaurant. "This is the way we are all doing," he said. Consulting the newspapers, I found that the first ship, the Nebraska, sailed for Aspinwall in about ten days, and that the steerage passage to

San Francisco by way of the Isthmus was only forty dollars.

In the mean time I wandered about the city without knowing a single person in it. My walks extended but little beyond sight of my little schooner home. I saw the name Central Park on some of the street-cars and thought I would like to visit it, but, fearing that I might not be able to find my way back, I dared not make the adventure. I felt completely lost in the vast throngs of people, the noise of the streets, and the immense size of the buildings. Often I thought I would like to explore the city if, like a lot of wild hills and valleys, it was clear of inhabitants.

The day before the sailing of the Panama ship I bought a pocket map of California and allowed myself to be persuaded to buy a dozen large maps, mounted on rollers, with a map of the world on one side and the United States on the other. In vain I said I had no use for them. "But surely you want to make money in California, don't you? Everything out there is very dear. We'll sell you a dozen of these fine maps for two dollars each and you can easily sell them in California for ten dollars apiece." I foolishly allowed myself to be persuaded. The maps made a very large, awkward bundle, but fortunately it was the only baggage I had except my little plant press and a small bag. I laid them in my berth in the steerage, for they were too large to be stolen and concealed.

There was a savage contrast between life in the steerage and my fine home on the little ship fruiter. Never before had I seen such a barbarous mob, especially at meals. Arrived at Aspinwall-Colon, we had half a day to ramble about before starting across the Isthmus. Never shall I forget the glorious flora, especially for the first fifteen or twenty miles along the Chagres River. The riotous exuberance of great forest trees, glowing in purple, red, and yellow flowers, far surpassed anything I had ever seen, especially of flowering trees, either in Florida or Cuba. I gazed from the car-platform enchanted. I fairly cried for joy and hoped that sometime I should be able to return and enjoy and study this most glorious of forests to my heart's content. We reached San Francisco about the first of April, and I remained there only one day, before starting for Yosemite Valley.[1]

I followed the Diablo foothills along the San José Valley to Gilroy, thence over the Diablo Mountains to the valley of the San Joaquin by the Pacheco Pass, thence down the valley opposite the mouth of the Merced River, thence across the San Joaquin, and up into the Sierra Nevada to the mammoth trees of Mariposa, and the glorious Yosemite, and thence down the Merced to this place.[2] The goodness of the weather as I journeyed toward Pacheco was beyond all praise and description — fragrant, mellow, and bright. The sky was perfectly delicious, sweet enough for the breath of angels; every draught of it gave a separate and distinct

[1] At this point the journal ends. The remainder of this chapter is taken from a letter written to Mrs. Ezra S. Carr from the neighborhood of Twenty Hill Hollow in July, 1868.

[2] Near Snelling, Merced County, California.

piece of pleasure. I do not believe that Adam and Eve ever tasted better in their balmiest nook.

The last of the Coast Range foothills were in near view all the way to Gilroy. Their union with the valley is by curves and slopes of inimitable beauty. They were robed with the greenest grass and richest light I ever beheld, and were colored and shaded with myriads of flowers of every hue, chiefly of purple and golden yellow. Hundreds of crystal rills joined song with the larks, filling all the valley with music like a sea, making it Eden from end to end.

The scenery, too, and all of nature in the Pass is fairly enchanting. Strange and beautiful mountain ferns are there, low in the dark cañons and high upon the rocky sunlit peaks; banks of blooming shrubs, and sprinklings and gatherings of garment flowers, precious and pure as ever enjoyed the sweets of a mountain home. And oh! what streams are there! beaming, glancing, each with music of its own, singing as they go, in shadow and light, onward upon their lovely, changing pathways to the sea. And hills rise over hills, and mountains over mountains, heaving, waving, swelling, in most glorious, overpowering, unreadable majesty.

When at last, stricken and faint like a crushed insect, you hope to escape from all the terrible grandeur of these mountain powers, other fountains, other oceans break forth before you; for there, in clear view, over heaps and rows of foothills, is laid a grand, smooth, outspread plain, watered by a river, and another range of peaky, snow-capped mountains a hundred miles in the distance. That plain is the valley of the San Joaquin, and those mountains are the great Sierra Nevada. The valley of the San Joaquin is the floweriest piece of world I ever walked, one vast, level, even flower-bed, a sheet of flowers, a smooth sea, ruffled a little in the middle by the tree fringing of the river and of smaller cross-streams here and there, from the mountains.

Florida is indeed a "land of flowers," but

for every flower creature that dwells in its most delightsome places more than a hundred are living here. Here, here is Florida! Here they are not sprinkled apart with grass between as on our prairies, but grasses are sprinkled among the flowers; not as in Cuba, flowers piled upon flowers, heaped and gathered into deep, glowing masses, but side by side, flower to flower, petal to petal, touching but not entwined, branches weaving past and past each other, yet free and separate — one smooth garment, mosses next the ground, grasses above, petaled flowers between.

Before studying the flowers of this valley and their sky, and all of the furniture and sounds and adornments of their home, one can scarce believe that their vast assemblies are permanent; but rather that, actuated by some plant purpose, they had convened from every plain and mountain and meadow of their kingdom, and that the different coloring of patches, acres, and miles marks the bounds of the various tribes and family encampments.

CHAPTER IX

TWENTY HILL HOLLOW [1]

WERE we to cross-cut the Sierra Nevada into blocks a dozen miles or so in thickness, each section would contain a Yosemite Valley and a river, together with a bright array of lakes and meadows, rocks and forests. The grandeur and inexhaustible beauty of each block would be so vast and over-satisfying that to choose among them would be like selecting slices of bread cut from the same loaf. One bread-slice might have burnt spots, answering to craters; another would be more browned; another, more crusted or raggedly cut; but all essentially the same. In no greater degree would the Sierra slices differ in general character. Nevertheless, we all would choose the Merced slice, because, being easier of access, it has been nibbled and tasted, and

[1] This is the hub of the region where Mr. Muir spent the greater part of the summer of 1868 and the spring of 1869.

pronounced very good; and because of the concentrated form of its Yosemite, caused by certain conditions of baking, yeasting, and glacier-frosting of this portion of the great Sierra loaf. In like manner, we readily perceive that the great central plain is one batch of bread—one golden cake—and we are loath to leave these magnificent loaves for crumbs, however good.

After our smoky sky has been washed in the rains of winter, the whole complex row of Sierras appears from the plain as a simple wall, slightly beveled, and colored in horizontal bands laid one above another, as if entirely composed of partially straightened rainbows. So, also, the plain seen from the mountains has the same simplicity of smooth surface, colored purple and yellow, like a patchwork of irised clouds. But when we descend to this smooth-furred sheet, we discover complexity in its physical conditions equal to that of the mountains, though less strongly marked. In particular, that portion of the plain lying between the

Merced and the Tuolumne, within ten miles of the slaty foothills, is most elaborately carved into valleys, hollows, and smooth undulations, and among them is laid the Merced Yosemite of the plain — Twenty Hill Hollow.

This delightful Hollow is less than a mile in length, and of just sufficient width to form a well-proportioned oval. It is situated about midway between the two rivers, and five miles from the Sierra foothills. Its banks are formed of twenty hemispherical hills; hence its name. They surround and enclose it on all sides, leaving only one narrow opening toward the southwest for the escape of its waters. The bottom of the Hollow is about two hundred feet below the level of the surrounding plain, and the tops of its hills are slightly below the general level. Here is no towering dome, no Tissiack, to mark its place; and one may ramble close upon its rim before he is made aware of its existence. Its twenty hills are as wonderfully regular in size and position as in form. They are like big marbles half buried in the

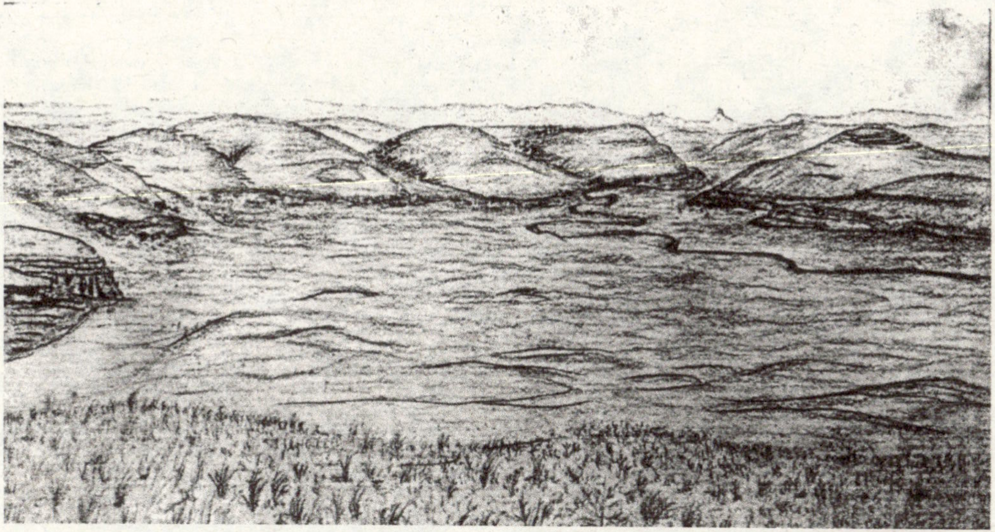

TWENTY HILL HOLLOW
From a sketch by Mr. Muir

ground, each poised and settled daintily into its place at a regular distance from its fellows, making a charming fairy-land of hills, with small, grassy valleys between, each valley having a tiny stream of its own, which leaps and sparkles out into the open hollow, uniting to form Hollow Creek.

Like all others in the immediate neighborhood, these twenty hills are composed of stratified lavas mixed with mountain drift in varying proportions. Some strata are almost wholly made up of volcanic matter — lava and cinders — thoroughly ground and mixed by the waters that deposited them; others are largely composed of slate and quartz boulders of all degrees of coarseness, forming conglomerates. A few clear, open sections occur, exposing an elaborate history of seas, and glaciers, and volcanic floods — chapters of cinders and ashes that picture dark days when these bright snowy mountains were clouded in smoke and rivered and laked with living fire. A fearful age, say mortals, when these Sierras flowed lava to the sea. What horizons of flame! What atmospheres of ashes and smoke!

The conglomerates and lavas of this region are readily denuded by water. In the time when their parent sea was removed to form this golden plain, their regular surface, in great part covered with shallow lakes, showed little variation from motionless level until torrents of rain and floods from the mountains gradually sculptured the simple page to the present diversity of bank and brae, creating, in the section between the Merced and the Tuolumne, Twenty Hill Hollow, Lily Hollow, and the lovely valleys of Cascade and Castle Creeks, with many others nameless and unknown, seen only by hunters and shepherds, sunk in the wide bosom of the plain, like undiscovered gold. Twenty Hill Hollow is a fine illustration of a valley created by erosion of water. Here are no Washington columns, no angular El Capitans. The hollow cañons, cut in soft lavas, are not so deep as to require a single earthquake at the hands of science, much less a baker's dozen of those convenient tools demanded for the making of mountain Yosemites, and our moderate arithmetical standards are not outraged by a single magnitude of this simple, comprehensible hollow.

The present rate of denudation of this portion of the plain seems to be about one tenth of an inch per year. This approximation is based upon observations made upon stream-banks and perennial plants. Rains and winds remove mountains without disturbing their plant or animal inhabitants. Hovering petrels, the fishes and floating plants of ocean, sink and rise in beautiful rhythm with its waves; and, in like manner, the birds and plants of the plain sink and rise with these waves of land, the only difference being that the fluctuations are more rapid in the one case than in the other.

In March and April the bottom of the Hollow and every one of its hills are smoothly covered and plushed with yellow and purple flowers, the yellow predominating. They are mostly social *Compositæ*, with a few claytonias, gilias, eschscholtzias, white and yellow violets, blue and yellow lilies, dodecatheons, and eriogonums set in a half-floating maze of purple grasses. There is but one vine in the Hollow — the *Megarrhiza* [*Echinocystis* T. & D.] or "Big Root." The only bush within a mile of it, about four feet in height, forms so remarkable an object upon the universal smoothness that my dog barks furiously around it, at a cautious distance, as if it were a bear. Some of the hills have rock ribs that are brightly colored with red and yellow lichens, and in moist nooks there are luxuriant mosses — *Bartramia*, *Dicranum*, *Funaria*, and several *Hypnums*. In cool, sunless coves the mosses are companioned with ferns — a *Cystopteris* and the little gold-dusted rock fern, *Gymnogramma triangularis*.

The Hollow is not rich in birds. The meadowlark homes there, and the little burrowing owl, the killdeer, and a species of sparrow. Occasionally a few ducks pay a visit to its waters, and a few tall herons — the blue and the white — may at times be seen stalking along the creek; and the sparrow hawk and gray eagle [1] come to hunt. The lark, who does nearly all the singing for the Hollow, is not identical in species with the meadowlark of the East, though closely resembling it; richer flowers and skies have inspired him with a better song than was ever known to the Atlantic lark.

I have noted three distinct lark-songs here. The words of the first, which I committed to memory at one of their special meetings, spelled as sung, are, "Wee-ro spee-ro wee-o weer-ly wee-it." On the 20th of January, 1869, they sang "Queed-lix boodle," repeating it with great regularity, for hours together, to music sweet as the sky that gave it. On the 22d of the same month, they sang "Chee chool chee-dildy choodildy." An inspiration is this song of the blessed lark, and universally absorbable by human souls. It seems to be the only bird-song of these hills that has been created with any direct reference to us. Music is one of the at-

[1] Mr. Muir doubtless meant the golden eagle (*Aquila chrysaëtos*).

tributes of matter, into whatever forms it may be organized. Drops and sprays of air are specialized, and made to plash and churn in the bosom of a lark, as infinitesimal portions of air plash and sing about the angles and hollows of sand-grains, as perfectly composed and predestined as the rejoicing anthems of worlds; but our senses are not fine enough to catch the tones. Fancy the waving, pulsing melody of the vast flower-congregations of the Hollow flowing from myriad voices of tuned petal and pistil, and heaps of sculptured pollen. Scarce one note is for us; nevertheless, God be thanked for this blessed instrument hid beneath the feathers of a lark.

The eagle does not dwell in the Hollow; he only floats there to hunt the long-eared hare. One day I saw a fine specimen alight upon a hillside. I was at first puzzled to know what power could fetch the sky-king down into the grass with the larks. Watching him attentively, I soon discovered the cause of his earthiness. He was hungry and stood watching a long-eared hare, which stood erect at the door of his burrow, staring his winged fellow mortal full in the face. They were about ten feet apart. Should the eagle attempt to snatch the hare, he would instantly disappear in the ground. Should long-ears, tired of inaction, venture to skim the hill to some neighboring burrow, the eagle would swoop above him and strike him dead with a blow of his pinions, bear him to some favorite rock table, satisfy his hunger, wipe off all marks of grossness, and go again to the sky.

Since antelopes have been driven away, the hare is the swiftest animal of the Hollow. When chased by a dog he will not seek a burrow, as when the eagle wings in sight, but skims wavily from hill to hill across connecting curves, swift and effortless as a bird-shadow. One that I measured was twelve inches in height at the shoulders. His body was eighteen inches, from nose-tip to tail. His great ears measured six and a half inches in length and two in width. His ears — which, notwithstanding their great size, he wears gracefully and becomingly — have procured for him the homely nickname, by which he is commonly known, of "Jackass rabbit." Hares are very abundant over all the plain and up in the sunny, lightly wooded foothills, but their range does not extend into the close pine forests.

Coyotes, or California wolves, are occasionally seen gliding about the Hollow, but they are not numerous, vast numbers having been slain by the traps and poisons of sheep-raisers. The coyote is about the size of a small shepherd-dog, beautiful and graceful in motion, with erect ears, and a bushy tail, like a fox. Inasmuch as he is fond of mutton, he is cordially detested by "sheep-men" and nearly all cultured people.

The ground-squirrel is the most common animal of the Hollow. In several hills there is a soft stratum in which they have tunneled their homes. It is interesting to observe these rodent towns in time of alarm. Their one circular street resounds with sharp, lancing outcries of

"Seekit, seek, seek, seekit!" Near neighbors, peeping cautiously half out of doors, engage in low, purring chat. Others, bolt upright on the doorsill or on the rock above, shout excitedly as if calling attention to the motions and aspects of the enemy. Like the wolf, this little animal is accursed, because of his relish for grain. What a pity that Nature should have made so many small mouths palated like our own!

All the seasons of the Hollow are warm and bright, and flowers bloom through the whole year. But the grand commencement of the annual genesis of plant and insect life is governed by the setting-in of the rains, in December or January. The air, hot and opaque, is then washed and cooled. Plant seeds, which for six months have lain on the ground dry as if garnered in a farmer's bin, at once unfold their treasured life. Flies hum their delicate tunes. Butterflies come from their coffins, like cotyledons from their husks. The network of dry water-courses, spread over valleys and hollows, suddenly gushes with bright waters, sparkling and pouring from pool to pool, like dusty mummies risen from the dead and set living and laughing with color and blood. The weather grows in beauty, like a flower. Its roots in the ground develop day-clusters a week or two in size, divided by and shaded in foliage of clouds; or round hours of ripe sunshine wave and spray in sky-shadows, like racemes of berries half hidden in leaves.

These months of so-called rainy season are not filled with rain. Nowhere else in North America, perhaps in the world, are Januarys so balmed and glowed with vital sunlight. Referring to my notes of 1868 and 1869, I find that the first heavy general rain of the season fell on the 18th of December. January yielded to the Hollow, during the day, only twenty hours of rain, which was divided among six rainy days. February had only three days on which rain fell, amounting to eighteen and one-half hours in all. March had five rainy days. April had three, yielding seven hours of rain.

May also had three wet days, yielding nine hours of rain, and completed the so-called "rainy season" for that year, which is probably about an average one. It must be remembered that this rain record has nothing to do with what fell in the night.

The ordinary rainstorm of this region has little of that outward pomp and sublimity of structure so characteristic of the storms of the Mississippi Valley. Nevertheless, we have experienced rainstorms out on these treeless plains, in nights of solid darkness, as impressively sublime as the noblest storms of the mountains. The wind, which in settled weather blows from the northwest, veers to the southeast; the sky curdles gradually and evenly to a grainless, seamless, homogeneous cloud; and then comes the rain, pouring steadily and often driven aslant by strong winds. In 1869, more than three fourths of the winter rains came from the southeast. One magnificent storm from the northwest occurred on the 21st of March; an immense, round-browed cloud came sailing over the flowery hills in most imposing majesty, bestowing water as from a sea. The passionate rain-gush lasted only about one minute, but was nevertheless the most magnificent cataract of the sky mountains that I ever beheld. A portion of calm sky toward the Sierras was brushed with thin, white cloud-tissue, upon which the rain-torrent showed to a great height — a cloud waterfall, which, like those of Yosemite, was neither spray, rain, nor solid water. In the same year the cloudiness of January, omitting rainy days, averaged 0.32; February, 0.13; March, 0.20; April, 0.10; May, 0.08. The greater portion of this cloudiness was gathered into a few days, leaving the others blocks of solid, universal sunshine in every chink and pore.

At the end of January, four plants were in flower: a small white cress, growing in large patches; a low-set, umbeled plant, with yellow flowers; an eriogonum, with flowers in leafless spangles; and a small boragewort. Five or six mosses had adjusted their hoods, and were in the prime of life. In February, squirrels, hares, and flowers were in springtime joy. Bright plant-constellations shone everywhere about the Hollow. Ants were getting ready for work, rubbing and sunning their limbs upon the husk-piles around their doors; fat, pollen-dusted, "burly, dozing humble-bees" were rumbling among the flowers; and spiders were busy mending up old webs, or weaving new ones. Flowers were born every day, and came gushing from the ground like gayly dressed children from a church. The bright air became daily more songful with fly-wings, and sweeter with breath of plants.

In March, plant-life is more than doubled. The little pioneer cress, by this time, goes to seed, wearing daintily embroidered silicles. Several claytonias appear; also, a large white leptosiphon[?], and two nemophilas. A small plantago becomes tall enough to wave and show silky ripples of shade. Toward the end of this month or the beginning of April, plant-life is at its greatest height. Few have any just conception of its amazing richness. Count the flowers of any portion of these twenty hills, or of the bottom of the Hollow, among the streams: you will find that there are from one to ten thousand upon every square yard, counting the heads of *Compositæ* as single flowers. Yellow *Compositæ* form by far the greater portion of this goldy-way. Well may the sun feed them with his richest light, for these shining sunlets are his very children — rays of his ray, beams of his beam! One would fancy that these California days receive more gold from the ground than they give to it. The earth has indeed become a sky; and the two cloudless skies, raying toward each other flower-beams and sunbeams, are fused and congolded into one glowing heaven. By the end of April most of the Hollow plants have ripened their seeds and died; but, undecayed, still assist the landscape with color from persistent involucres and corolla-like heads of chaffy scales.

In May, only a few deep-set lilies and eriogonums are left alive. June, July, August, and

September are the season of plant rest, followed, in October, by a most extraordinary outgush of plant-life, at the very driest time of the whole year. A small, unobtrusive plant, *Hemizonia virgata,* from six inches to three feet in height, with pale, glandular leaves, suddenly bursts into bloom, in patches miles in extent, like a resurrection of the gold of April. I have counted upward of three thousand heads upon one plant. Both leaves and pedicles are so small as to be nearly invisible among so vast a number of daisy golden-heads that seem to keep their places unsupported, like stars in the sky. The heads are about five eighths of an inch in diameter; rays and disk-flowers, yellow; stamens, purple. The rays have a rich, furred appearance, like the petals of garden pansies. The prevailing summer wind makes all the heads turn to the southeast. The waxy secretion of its leaves and involucres has suggested its grim name of "tarweed," by which it is generally known. In our estimation, it is the most delightful member of the whole Composite Family of the plain. It remains in flower until November, uniting with an eriogonum that continues the floral chain across December to the spring plants of January. Thus, although nearly all of the year's plant-life is crowded into February, March, and April, the flower circle around the Twenty Hill Hollow is never broken.

The Hollow may easily be visited by tourists *en route* for Yosemite, as it is distant only about six miles from Snelling's. It is at all seasons interesting to the naturalist; but it has little that would interest the majority of tourists earlier than January or later than April. If you wish to see how much of light, life, and joy can be got into a January, go to this blessed Hollow. If you wish to see a plant-resurrection,—myriads of bright flowers crowding from the ground, like souls to a judgment,—go to Twenty Hills in February. If you are traveling for health, play truant to doctors and friends, fill your pocket with biscuits, and hide in the hills of the Hollow, lave in its waters, tan in its golds, bask in its flower-shine, and your baptisms will make you a new creature indeed. Or, choked in the sediments of society, so tired of the world, here will your hard doubts disappear, your carnal incrustations melt off, and your soul breathe deep and free in God's shoreless atmosphere of beauty and love.

Never shall I forget my baptism in this font. It happened in January, a resurrection day for many a plant and for me. I suddenly found myself on one of its hills; the Hollow overflowed with light, as a fountain, and only small, sunless nooks were kept for mosseries and ferneries. Hollow Creek spangled and mazed like a river. The ground steamed with fragrance. Light, of unspeakable richness, was brooding the flowers. Truly, said I, is California the Golden State — in metallic gold, in sun gold, and in plant gold. The sunshine for a whole summer seemed condensed into the chambers of that one glowing day. Every trace of dimness had been washed from the sky; the mountains were dusted and wiped clean with clouds — Pacheco Peak and Mount Diablo, and the waved blue wall between; the grand Sierra stood along the plain, colored in four horizontal bands:— the lowest, rose purple; the next higher, dark purple; the next, blue; and, above all, the white row of summits pointing to the heavens.

It may be asked, What have mountains fifty or a hundred miles away to do with Twenty Hill Hollow? To lovers of the wild, these mountains are not a hundred miles away. Their spiritual power and the goodness of the sky make them near, as a circle of friends. They rise as a portion of the hilled walls of the Hollow. You cannot feel yourself out of doors; plain, sky, and mountains ray beauty which you feel. You bathe in these spirit-beams, turning round and round, as if warming at a camp-fire. Presently you lose consciousness of your own separate existence: you blend with the landscape, and become part and parcel of nature.

THE END

Index

Index

Index

Index

Index

Index